W0043955

D. J. Gardiner P. R. Graves (Editors)

Practical Raman Spectroscopy

With Contributions by
H. J. Bowley, D. J. Gardiner, D. L. Gerrard,
P. R. Graves, J. D. Louden, G. Turrell

With 87 Figures and 11 Tables

Springer-Verlag Berlin Heidelberg NewYork
London Paris Tokyo

Editors

Dr. Derek J. Gardiner

Raman Applications and Instrument Development Group, Department of Chemical and Life Sciences, Newcastle upon Tyne Polytechnic, Newcastle upon Tyne, NE1 8ST, England

Dr. Pierre R. Graves

Harwell Laboratory, Materials Development Division, B. 393
UK Atomic Energy Authority, Oxfordshire OX11 ORA, England

Co-Authors

Dr. Heather J. Bowley

BP Research Centre, Chertsey Road, Sunbury on Thames, Middlesex TW16 7LN, England

Dr. Don L. Gerrard

BP Research Centre, Chertsey Road, Sunbury on Thames, Middlesex TW16 7LN, England

Dr. John D. Louden

I.C.I. Mond Division, P.O. Box No. 8, The Heath, Runcorn, Cheshire, England

Professor G. Turrell

CNRS, Université des Sciences et Techniques de Lille, Laboratoire
de Spectroscopie Infrarouge et Raman, 59655 Villeneuve-D'ascq, France

ISBN-13: 978-3-540-50254-8   e-ISBN-13: 978-3-642-74040-4
DOI: 10.1007/978-3-642-74040-4

Library of Congress Cataloging in Publication Data.
Practical Raman spectroscopy / [edited by] D. J. Gardiner, P. R. Graves:
with contributions by H. J. Bowley . . . [et al.].
p. cm. Bibliography: p.
                1. Raman spectroscopy. I. Gardiner, D. J. (Derek J.), 1945– .
II. Graves, P. R. (Pierre R.) III. Bowley, H. J. (Heather J.)
QD96.R34P73 1988   543'.08584--dc19   88-30444 CIP

This work is subject to copyright. All rights are reserved, whether the whole of part of the material is concerned, specifically the rights of translation, reprinting, reuse of illustrations, recitation, broadcasting, reproduction on microfilms or in other ways, and storage in data banks. Duplication of this publication or parts thereof is only permitted under the provisions of the German Copyright Law of September 9, 1965, in its version of June 24, 1985, and a copyright fee must always be paid. Violations fall under the prosecution act of the German Copyright Law.

© Springer-Verlag Berlin Heidelberg 1989

The use of registered names, trademarks, etc. in this publication does not imply, even in the absence of a specific statement, that such names are exempt from the relevant protective laws and regulations and therefore free for general use.

2152/3020-543210 — Printed on acid-free paper

Preface

This volume sets out to draw together the essential expertise which will provide a technical guide to the practice of Raman spectroscopy. The text deals exclusively with spontaneous Raman spectroscopy and includes some aspects of Resonance Raman spectroscopy. Chapter 1 sets out the essential theoretical framework using a simple classical approach and deals with the rudiments of polarizability. Many of these theoretical points are further developed in Chap. 2 where the scattering and polarization consequences of various sampling geometries and collection optics, on gaseous, liquid, single crystal and thin film methods are detailed. The relative advantages and disadvantages of the wide variety of hardware now available to the Raman spectroscopist are discussed in Chap. 3. Important calibration data is presented in Chap. 4 along with an account of data analysis techniques, including signal enhancement methods. Chapter 5 describes some of the techniques and cell designs that have been successfully used to study samples under extreme conditions and Chap. 6 deals with the rapidly growing technique of Raman microscopy, providing a wide range of application examples and experimental advice.

We recognise the difficulty in covering all aspects of Raman spectroscopy in a single volume and a section on further reading, representing what we feel are amongst the more informative references, at the time of publication, is provided for additional detail.

Our hope is that *Practical Raman Spectroscopy* will help to provide a source of on-hand technical support and data for the practising Raman spectroscopist in the laboratory.

February 1989   Derek J. Gardiner and Pierre R. Graves

Contents

CHAPTER 1

Introduction to Raman Scattering

by Derek J. Gardiner
Raman Applications and Instrument Development Group
Department of Chemical and Life Sciences, Newcastle upon
Tyne Polytechnic, Newcastle upon Tyne, UK, NE1 8ST

1.1 Historical Introduction

The Raman effect was predicted by Smekal in 1923 but was first observed by Raman in 1928. The first experiments were carried out using focussed sunlight and filters and relied on visual observation of colour changes in the scattered light. Later Raman recorded spectra of several liquids including benzene and carbontetrachloride using a mercury lamp and a spectrograph. The experiments used 600 ml of sample and required a 24 hour exposure to obtain measurable spectra. The Raman effect is an inherently weak effect, typically 10^{-8} of the intensity of the incident exciting radiation and for many years source stability and intensity made Raman spectroscopy extremely difficult particularly in comparison with the faster and less expensive infrared spectrometers that were developed. In 1952, a major improvement occurred with the introduction of the Toronto arc source. This comprised a four turn helix of pyrex glass capable of radiating as much as 50 watt in the 435.8 nm emission line of mercury, though only a fraction of this power could be used practically to excite a spectrum.

It was not until the early sixties that the modern Raman renaissance took place with the development of commercial CW visible lasers. Suddenly a highly monochromatic, coherent, narrow beam, high intensity light source was available which revolutionised Raman spectroscopy. Spectra could easily be recorded from very small sample volumes, coloured samples, solids, liquids and gases and samples at high temperature, in dilute solutions, under vacuum and in a variety of other nonstandard conditions.

In recent years, microelectronics has further improved the technique such that stepper motor drives, photon counting, digital data aquisition and computer processing have provided chemists, physicists and analysts with a technique which many claim is more useful and versatile than infrared spectroscopy. The speed with which a Raman spectrum can be recorded depends upon the response of the detection

Practical Raman Spectroscopy, Gardiner and Graves (Eds.)
ⓒ Springer-Verlag Berlin Heidelberg 1989

system and by using optical multichannel analysers to detect the spread of dispersed wavelengths, Raman spectra can be recorded in microseconds. This is not only an aid to routine analysis but opens up the possibility of time resolved spectroscopy and kinetic studies. Along with these improvements in technique have come the discovery and observation of related spectroscopic effects such as the stimulated Raman effect, the hyper Raman effect, Raman gain spectroscopy, Raman induced Kerr effect spectroscopy and the induced Raman effect. Resonance Raman spectroscopy has allowed the study of chromophoric species at very low concentrations and now plays an important role in the study of biologically significant molecules. In recent years, the study of high temperature gases and flames has been greatly advanced by the use of coherent anti-Stokes Raman spectroscopy.

Raman spectroscopy and its related techniques are now finding application across a wide range of research activity and are increasingly becoming standard techniques in academic and industrial analytical laboratories.

1.2 Spontaneous Raman Theory

There exist extensive, detailed descriptions of the development of Raman theory. The approach taken here is intended as a simplified outline and contains many of the significant relationships sufficient to understand the essential nature of the effect. Approaches used by others in reviewing this material have been used, principally those of Gilson and Hendra [1], Chantry [2], Clark [3], Long [4] and Szymanski [5]. Further details are contained in these reviews and references therein.

Simply stated, the Raman effect can be described as the inelastic scattering of light by matter. When a photon of visible light, too low in energy to excite an electronic transition, interacts with a molecule it can be scattered in one of three ways. It can be elastically scattered and thus retain its incident energy or it can be inelastically scattered by either giving energy up to, or by removing energy from, the molecule. Photons undergoing inelastic loss of energy give rise to Stokes scattering whilst photons undergoing inelastic gain of energy give rise to anti-Stokes scattering. The energy gained by the molecule in Stokes scattering appears as vibrational energy and where a molecule has excess vibrational energy above the ground state, it is this energy which is lost to the anti-Stokes scattered photons.

1.2.1 Classical Theory

The classical approach to a description of the Raman effect regards the scattering molecule as a collection of atoms undergoing simple harmonic vibrations and takes no account of quantisation of the vibrational energy.

When a molecule is placed in an electric field, its electrons are displaced relative to its nuclei thus developing an electric dipole moment. For small fields, the induced dipole moment μ_i is proportional to the field strength ε.

$$\mu_i = \alpha\varepsilon \qquad (1)$$

The proportionality constant α is the polarisability of the molecule, that is the ease with which the electron cloud of the molecule can be distorted. A fluctuating electric field will produce a fluctuating dipole moment of the same frequency. Electromagnetic radiation generates such an electric field which can be expressed as:

$$\varepsilon = \varepsilon^0 \, Cos \, 2\pi t v_0 \tag{2}$$

ε^0 is the equilibrium field strength and v_0 is the angular frequency of the radiation. Thus electromagnetic radiation will induce a fluctuating dipole of frequency v_0 in the molecule. This induced dipole will emit or scatter radiation of frequency v_0. This is Rayleigh scattering.

Consider the particular case of a diatomic molecule which vibrates with a frequency of v_v. If it performs simple harmonic vibrations, then a coordinate q_v along the axis of vibration at time t, is given by:

$$q_v = q_0 \, Cos \, 2\pi v_v t \tag{3}$$

If the polarisability changes during the vibration, its value for a small vibrational amplitude will be given by:

$$\alpha = \alpha^0 + \left(\frac{\partial \alpha}{\partial q_v} \right)_0 q_v \tag{4}$$

Substitution of Eq. (3) in Eq. (4) yields:

$$\alpha = \alpha^0 + \left(\frac{\partial \alpha}{\partial q_v} \right)_0 q_0 \, Cos \, 2\pi v_v t \tag{5}$$

If incident radiation of frequency v_0 interacts with the molecule then from Eq. (1) and (2):

$$\mu_i = \alpha \varepsilon = \alpha \varepsilon^0 \, Cos \, 2\pi v_0 t \tag{6}$$

substitution of Eq. (5) in Eq. (6) yields:

$$\mu_i = \alpha^0 \varepsilon^0 \, Cos \, 2\pi v_0 t + \left(\frac{\partial \alpha}{\partial q_v} \right)_0 \varepsilon^0 q_0 \, Cos \, 2\pi v_v t \, Cos \, 2\pi v_0 t$$

which can be rewritten as:

$$\mu_i = \alpha^0 F^0 \, Cos \, 2\pi v_0 t + \left(\frac{\partial \alpha}{\partial q_v} \right)_0 \frac{\varepsilon^0 q_0}{2}$$

$$\times \left[Cos \, 2\pi (v_0 + v_v) \, t + Cos \, 2\pi (v_0 - v_v) \, t \right] \tag{7}$$

The first term in Eq. (7) describes the Rayleigh scattering and the remaining terms describe the Stokes and the anti-Stokes Raman scatter. Equation (7) indicates that light will be scattered with frequencies:

$$v_0 = \text{RAYLEIGH SCATTER}$$

and

$$v_0 \pm v_v = \text{RAMAN SCATTER}$$

In addition Eq. (7) shows that for Raman scattering to occur:

$$\left(\frac{\partial \alpha}{\partial q_v}\right)_0 \neq 0 \tag{8}$$

that is, the polarisability of the molecule must change during a vibration if that vibration is to be Raman active.

1.2.2 Quantum Theory

The quantum theory approach to Raman scattering recognises that the vibrational energy of a molecule is quantised. A non-linear molecule will have 3N-6 normal vibrations and a linear molecule will have 3N-5, where N is the number of atoms in the molecule. The energy of each of these vibrations will be quantised according to the relationship:

$$E_v = hv(v + 1/2) \tag{9}$$

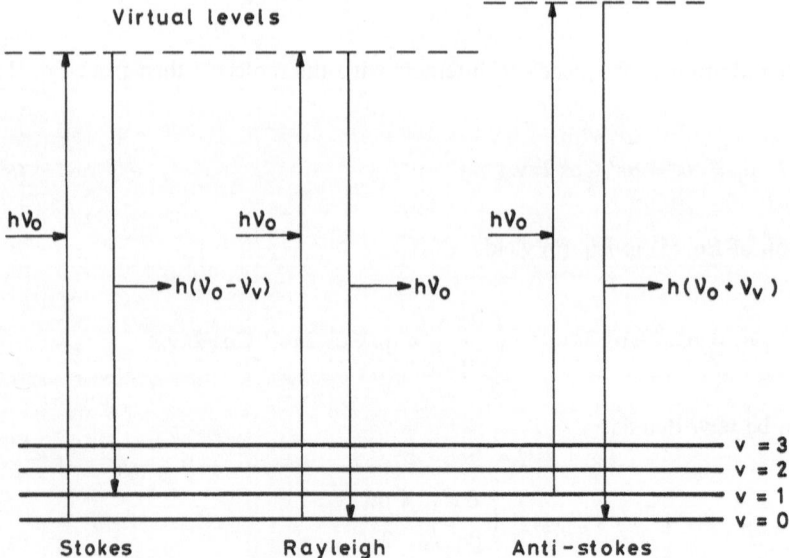

Fig. 1.1. Idealised model of Rayleigh scattering and Stokes and anti-Stokes Raman scattering

where v is the frequency of the vibration and v is the vibrational quantum number controlling the energy of that particular vibration and having values of 0, 1, 2, 3, ... etc. Perturbation theory is used to introduce quantisation into the Raman scattering theory. Put simply this approach applies perturbations to the ground state molecular wavefunctions until new wavefunctions are obtained which describe the vibrational excited state. The transition from ground state can then be regarded as being achieved via a perturbing wavefunction which is the sum of the perturbations applied. This perturbing wavefunction will have a corresponding energy and gives us a useful pictorial description of Raman scattering with the vibrational transitions ocurring via this virtual energy level (Fig. 1.1).

The Rayleigh scattering arises from transitions which start and finish at the same vibrational energy level. Stokes Raman scattering arises from transitions which start at the ground state vibrational energy level and finish at a higher vibrational energy level, whereas anti-Stokes Raman scattering involves a transition from a higher to a lower vibrational energy level. At normal room temperatures, most molecular vibrations are in the ground, v = 0 state and thus the anti-Stokes transitions are less likely to occur than the Stokes transitions resulting in the Stokes Raman scattering being more intense. This greater relative intensity becomes increasingly greater as the energy of the vibrations increases and the higher vibrational energy levels become less populated at any given temperature. For this reason it is usually

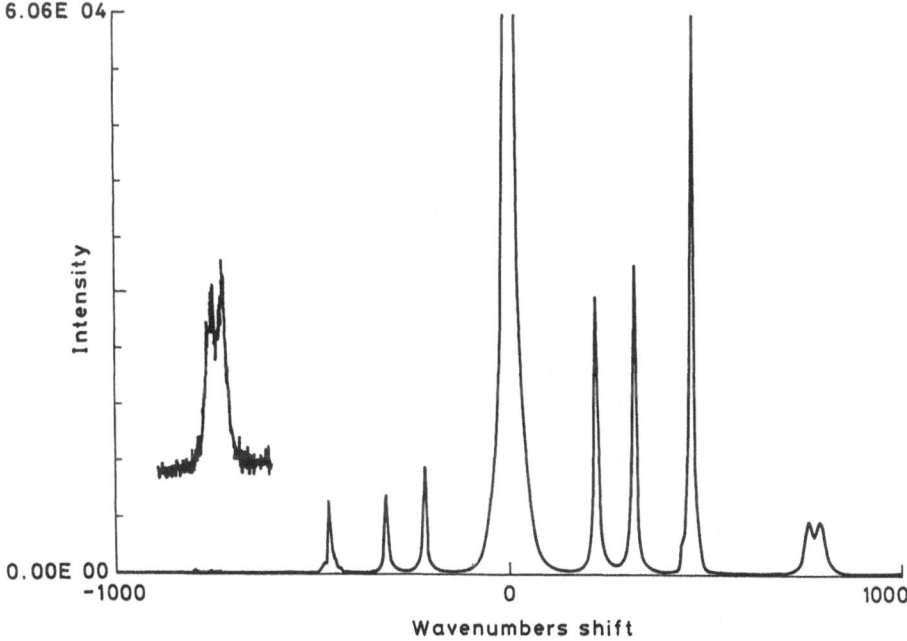

Fig. 1.2. Stokes and anti-Stokes Raman spectrum of carbontetrachloride obtained with a Spex 1403 monochromator, using 90° scattering from a single pass through a sample held in a capilliary tube. Excitation was ~200 mW of 514.5 nm radiation from an Ar^+ laser and the scan parameters were 1 cm^{-1} step, 0.5 s integration and 2 cm^{-1} bandpass. The anti-Stokes band at ~ −790 cm^{-1} has been magnified by X100.

the Stokes Raman scattering which is routinely studied and implied in Raman spectroscopy. This intensity effect can clearly be seen in Fig. 1.2 where the complete Stokes and anti-Stokes Raman spectrum of carbon tetrachloride is shown.

1.2.3 Polarisability

Polarisability is defined in section 1.2.1 and Eq. (1).

$$\mu_i = \alpha\varepsilon$$

The polarisability is a tensor quantity and as such has components in each of the x, y, z directions of a cartesian coordinate system. As a result, an electric field component in one direction can induce a dipole component in the x, y and z directions as defined by the following equations:

$$\mu_x = \alpha_{xx}\varepsilon_x + \alpha_{xy}\varepsilon_y + \alpha_{xz}\varepsilon_z$$
$$\mu_y = \alpha_{yx}\varepsilon_x + \alpha_{yy}\varepsilon_y + \alpha_{yz}\varepsilon_z$$
$$\mu_z = \alpha_{zx}\varepsilon_x + \alpha_{zy}\varepsilon_y + \alpha_{zz}\varepsilon_z$$

which can be written in matrix notation as:

$$\begin{pmatrix} \mu_x \\ \mu_y \\ \mu_z \end{pmatrix} = \begin{pmatrix} \alpha_{xx} & \alpha_{xy} & \alpha_{xz} \\ \alpha_{yx} & \alpha_{yy} & \alpha_{yz} \\ \alpha_{zx} & \alpha_{zy} & \alpha_{zz} \end{pmatrix} \begin{pmatrix} \varepsilon_x \\ \varepsilon_y \\ \varepsilon_z \end{pmatrix} \tag{10}$$

or

$$\mu = \alpha\varepsilon \tag{11}$$

where α is the polarisability tensor.

Normally $\alpha_{ij} = \alpha_{ji}$ and thus there are six independent components of this two dimensional tensor. The quantum mechanical formulation leads to the same result except that a Raman tensor is derived with components R_{ij}.

Changes in any $\alpha_{ij}(R_{ij})$ result in Raman activity. The components $\alpha_{ij}(R_{ij})$ can be used to generate irreducible representations of the molecular point group. It then follows that the $\alpha_{ij}(R_{ij})$ components are only changed by vibrations which generate the same irreducible representation, i.e. have the same symmetry. Using group theory in this way the Raman activity of the various vibrational modes of a molecule can be predicted.

The treatment of polarisability developed so far can usefully be applied to scattering by fluids. There are two invariant (constant regardless of the orientation of the molecule) properties of the polarisability tensor:

(1) the mean value $\bar{\alpha}$
(2) the anisotropy γ.

The mean value can be written as:

$$\bar{\alpha} = 1/3(\alpha_{xx} + \alpha_{yy} + \alpha_{zz}) \tag{12}$$

and the anisotropy as:

$$\gamma = 1/2 \left[(\alpha_{xx} - \alpha_{yy})^2 + (\alpha_{yy} - \alpha_{zz})^2 + (\alpha_{zz} - \alpha_{xx})^2 \right.$$
$$\left. + 6 \, (\alpha_{xy}^2 + \alpha_{xz}^2 + \alpha_{yz}^2) \right] \tag{13}$$

One consequence of the polarisability tensor being symmetrical is that spontaneously scattered light with directional properties is produced. These can be expressed in terms of an *average* scattering tensor for the assembly of randomly orientated tumbling molecules if it contains only the invariant parts of the individual molecular polarisability tensors. Thus the *average* scattering tensor is written with:

$$\overline{\alpha_{ii}^2} = \overline{\alpha_{xx}^2} = \overline{\alpha_{yy}^2} = \overline{\alpha_{zz}^2} = \frac{1}{45} \, (45\bar{\alpha}^2 + 4\gamma^2) \tag{14}$$

and

$$\overline{\alpha_{ij}^2} = \overline{\alpha_{xy}^2} = \overline{\alpha_{xz}^2} = \overline{\alpha_{yz}^2} = \frac{1}{15} \, \gamma^2 . \tag{15}$$

As the components of the *average* scattering tensor are invariant the x, y, z coordinates can be chosen to coincide with any convenient laboratory axes. Essentially the complex tumbling assembly of molecules is regarded as being represented by one *average* molecule orientated according to the laboratory axes and where $\alpha = \bar{\alpha}$.

Two types of directional scatter can be identified.

(1) *Isotropic scatter*: this is governed by the $\overline{\alpha_{ii}^2}$ terms and is scattered by dipoles orientated in the same direction as the electric vector of the exciting radiation.

(2) *Anisotropic scatter*: this is governed by the $\overline{\alpha_{ij}^2}$ terms and is scattered by dipoles orientated in the plane perpendicular to the electric vector of the exciting radiation.

The most common scattering geometry is 90°, scattering and uses the laboratory axes indicated in Fig. 1.3.

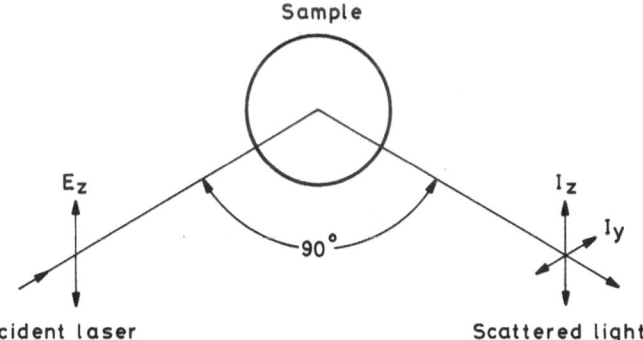

Fig. 1.3. 90° scattering geometry

Intensity of the scattered light is proportional to the square of the induced dipole moment and thus for the isotropic and anisotropic scattering; I_z and I_y in Fig. 1.3 we can write:

$$I_z = k\overline{\alpha_{ii}^2}\varepsilon_z^2$$
$$I_y = \overline{k\alpha_{ij}^2}\varepsilon_z^2 .$$

The ratio of the intensities of the anisotropic and isotropic scatter is a useful quantity refered to as the *depolarisation ratio* ϱ.

$$\varrho = \frac{I_y}{I_z} = \frac{\overline{\alpha_{ij}^2}}{\overline{\alpha_{ii}^2}} = \frac{1/15\gamma^2}{1/45(45\alpha^2 \ 4\gamma^2)} = \frac{3\gamma^2}{45\alpha^2 + 4\gamma^2} . \tag{16}$$

The quantum theory treatment of depolarisation ratios is analogous except that we write R for α. For a non-totally symmetrical vibration, that is a vibration which does not generate the totally symmetrical irreducible representation of the molecular point group, $R = 0$ and thus $\varrho = 3/4$. For all other vibrations $0 < \varrho < 3/4$. Using this result the value of the depolarisation ratio enables the symmetry of a vibration to be determined.

1.2.4 Raman intensities

A considerable amount of effort has been devoted to the derivation of realistic quantum mechanical expressions for Raman intensities. It is not appropriate here to discuss the relative merits of these and thus a simpler semi-classical account will be used based upon Placzek's polarisability theory, which is entirely adequate for a basic understanding of the factors involved. The theory applies to non-rotating molecules with a singlet ground state which conform to the Born-Oppenheimer approximation and involves exciting light frequencies that are very much less than any electronic frequency of the molecule. This set of conditions in fact applies quite well to the large majority of molecules. The total intensity of a Stokes Raman band of shift frequency v, scattered over a solid angle of 4π by a randomly oriented molecule perturbed by electromagnetic radiation from an initial vibronic state m to a final state n is given by:

$$I_{mn} = \frac{2^7\pi^5}{3^2C^4} I_0(v_0 - v)^4 \sum_{ij} |(\alpha_{ij})_{mn}|^2 \tag{17}$$

where I_0 is the incident intensity and α_{ij} represents the components of the polarisability tensor associated with the transition m → n.

The components α_{ij} can be expanded as a function of vibrational coordinate (Q_{mn}) in a Taylor series:

$$[\alpha_{ij}]_{mn} = \alpha_0 \int \gamma_m\gamma_n \ dt + \frac{\partial\alpha}{\partial Q_{mn}} \int \gamma^* Q_{mn}\gamma_n \ dt . \tag{18}$$

The first term here is responsible for Rayleigh and the second term for Raman scattering. If this second term is combined with Eq. (17) an expression for intensity arising from N molecules is obtained.

$$I_{mn} = \frac{2^4\pi^3}{3^2C^4} \cdot \frac{hI_0N(\nu_0 - \nu)^4}{\mu\nu(1 - e^{-h\nu/kT})} \sum_{ij} \left(\frac{\partial\alpha_{ij}}{\partial Q}\right)^2 \tag{19}$$

Carrying out the summation over ij using Eq. (14) and (15) and collecting all the constants together as K we obtain:

$$I_{mn} = \frac{KI_0N(\nu_0 - \nu)^4}{\mu\nu(1 - e^{-h\nu/kT})} [45\alpha^2 + 7\gamma^2] \tag{20}$$

where μ is the reduced mass of the oscillator. This expression can be further modified depending upon the scattering geometry via Eq. (16) to yield an expression which involves the depolarisation ratio.

1.3 Resonance Raman Theory

1.3.1 The Resonance Condition

Equation (17) indicates the way in which the intensity of a Raman band is related to the polarisability tensor. Using third order time dependent perturbation theory the following expression for the polarisability tensor can be derived which involves the electronic states of the molecule.

$$\alpha_{ij} = \frac{1}{h} \sum_{ef} \frac{(M_j)_{ge} \, h_{ef}^{\Delta\nu}(M_i)_{fg}}{(\nu_e - \nu_0)(\nu_f - \nu_s)} + \frac{(M_i)_{ge} \, h_{ef}^{\Delta\nu}(M_j)_{fg}}{(\nu_e + \nu_s)(\nu_f + \nu_0)} \tag{21}$$

Here the summation is taken over all excited states of the molecule e and f taken in pairs. The electronic ground state is g and M_j and M_i are the electric dipole transition moments in the directions j and i between the levels indicated. The term $h_{ef}^{\Delta\nu}$ is a vibronic coupling term connecting the states e and f by the vibration v and ν_0 is the frequency of the exciting radiation. The damping terms which prevent the denominator going to zero as ν_0 approaches ν_e have been omitted. It is clear from Eq. (21) that as the frequency of the exciting radiation ν_0 approaches ν_e then the value of the polarisability tensor components will increase rapidly. This is the resonance condition and results in increases in Raman intensity of several orders of magnitude. Two broad types of resonance effect can be identified; the pre-resonance Raman effect (pre-RRE) and the resonance Raman effect (RRE). Typically the pre-RRE is observed when the exciting radiation frequency comes within the high or low frequency wings but not under the observable vibrational structure of the electronic absorption band involved in the Raman scattering process. When the exciting radiation frequency falls within the observable vibrational structure, then the RRE is observed.

1.3.2 Intensity and Exciting Frequency

Two general catagories of frequency dependence can be identified corresponding to a resonance condition in which only one electronic level is important (e = f in Eq. (21)) and the case where two levels are involved (e ≠ f). In the first case which applies only to wholly symmetric vibrational modes:

$$I_i \alpha (v_0 - v_i)^4 \left[\frac{(v_e^2 + v_0^2)}{(v_e^2 - v_0^2)^2} \right]^2 = F_A^2 \tag{22}$$

and in the second case, which applies to vibrational modes having any symmetry contained in the direct product of the representations of the two electronic states involved:

$$I_i \alpha 4(v_0 - v_i)^4 \left[\frac{(v_e v_s + v_0^2)}{(v_e^2 - v_0^2)(v_f^2 - v_0^2)} \right]^2 = F_B^2 \tag{23}$$

In most systems it appears that at least two electronic states are involved in the resonance process and thus the expression F_B^2 is more widely applicable.

The vibrational modes most enhanced in the great majority of cases are wholly symmetrical stretching vibrations. In addition it is common to observe extensive overtone progressions for the more resonance enhanced modes. For simple molecules up to 10 or 20 overtones have been observed.

1.3.3 Excitation Profiles

A considerable amount of work has been published exploring the relationship between the extent of enhancement and the nature of the vibrational modes. It is clear that resonance enhancement is greatest for those vibrational modes which are involved in the vibronic structure of the absorption band. Plots of Raman intensity corrected for spectral response of the spectrometer and for v^4, preferably using an internal standard, against excitation frequency, are refered to as excitation profiles. Such a plot for a vibrational mode which is coupled to the electronic transition will essentially duplicate the absorption band profile. In this way valuable information relating to the assignment and vibrational structure of electronic spectra can be obtained. Figure 1.4 shows excitation profiles for the permanganate ion in $KMnO_4/KClO_4$ mixed crystal obtained by measuring the resonance enhancement of the intensity of v_1 and the overtones $2v_1$ and $3v_1$ as a function of excitation wavelength. Best calculated fits to the data allow the change in equilibrium bond length arising from the $^1T_2 \leftarrow {}^1A_1$ transition to be estimated [6].

1.3.4 Depolarisation

Theoretically, for free molecules belonging to non-cubic point groups, possessing a threefold or higher axis and having a non-degenerate electronic ground state, the pre-RRE allows determination of the symmetry of the electronic state which is in-

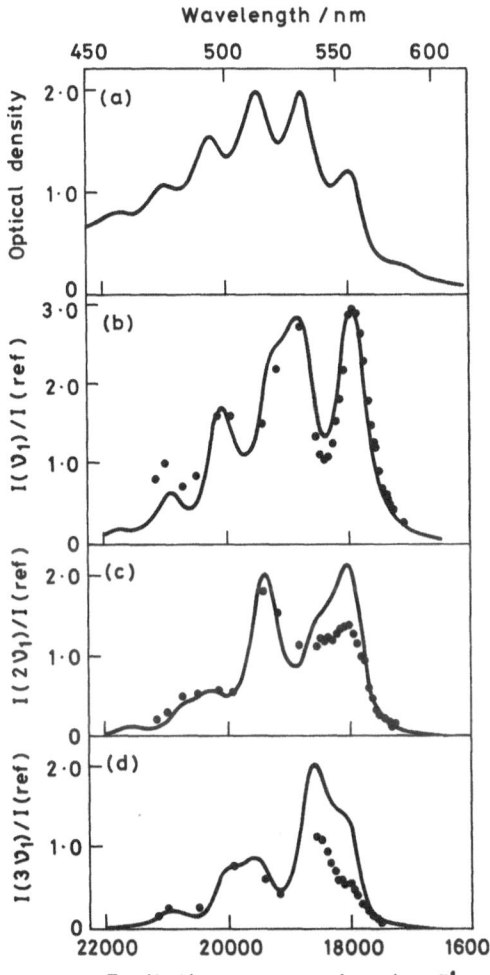

Fig. 1.4. a. Absorption spectrum of K[MnO$_4$]/K[ClO$_4$] mixed crystal in a K[ClO$_4$] disk at room temperature. b–d Experimental excitation profiles (●) for v_1, $2v_1$ and $3v_1$ bands respectively, of K[MnO$_4$]/K[ClO$_4$] mixed crystal. The full lines represent the best fit calculated profiles. [Reproduced with permission from reference 6]

volved in the resonance. The depolarisation ratio, defined in Eq. (16), is predicted to have a value of 1/3 for a band involving a non-degenerate wholly symmetrical electronic state and 1/8 for a band which involves a doubly degenerate state. For molecules belonging to the cubic point groups, which involve a triply degenerate electronic state, the depolarisation ratio is zero.

For a non-totally symmetric vibrational mode for which the Raman scattering tensor is antisymmetric (see Sect. 1.2.3) inverse depolarisation can occur in which $\varrho = \infty$. This effect can only be observed under resonance conditions as otherwise these vibrations are inactive.

1.4 References

1. Gilson TR, Hendra PJ (1972) Laser Raman spectroscopy. Wiley-Interscience, London
2. Chantry GW (1971) The Raman effect Volume 1 Ch 2. Marcel Dekker, New York
3. Clark RJH (1975) Advances Infrared and Raman Spectrosc. 1: 143
4. Long DA (1977) Raman spectroscopy. McGraw-Hill, UK
5. Woodward LA (1967) General introduction. In: Szymanski HA (ed) Raman spectroscopy theory and practice. Plenum, New York, Ch 1
6. Clark RJH, Stewart B (1981) J. Am. Chem. Soc. 103: 6593

CHAPTER 2

Raman Sampling

by George Turrell
Laboratoire de Spectrochimie Infrarouge et Raman (LP 2641, CNRS)
Université des Sciences et Techniques de Lille-Flandres-Artois
59655 Villeneuve d'Ascq, France

2.1 Introduction

Following Placzek [1] the intensity of Raman scattering can be expressed in the form

$$\mathscr{I}_s = I_e K \, |\tilde{e}_e \alpha e_s|^2 \, d\Omega \,, \tag{1}$$

where $K = 4\pi^2 a^2 \bar{v}_s^4$, $a \simeq 1/137$ and $\bar{v}_s$ is the wavenumber of the scattered light. The scattered energy per unit time (intensity) into a solid angle $d\Omega$ is given by $\mathscr{I}_s$, while I_e is the energy per unit area per unit time (irradiance) of the excitation incident on the sample. The unit vectors e_e and e_s define the directions of the electric fields of the exciting and scattered radiation, respectively, and α is the scattering tensor.

In the classical theory of the Raman effect the dipole moment induced in the sample is given by

$$\mu_i = \alpha \mathscr{E}_e \,, \tag{2}$$

where $\mathscr{E}_e$ is the electric field of the incident radiation and the polarizability tensor α is analogous to the scattering tensor introduced in Eq. (1). The linear transformation given by Eq. (2) is correct in the limit of weak radiation, in which case higher order

Practical Raman Spectroscopy, Gardiner and Graves (Eds.)
(c) Springer-Verlag Berlin Heidelberg 1989

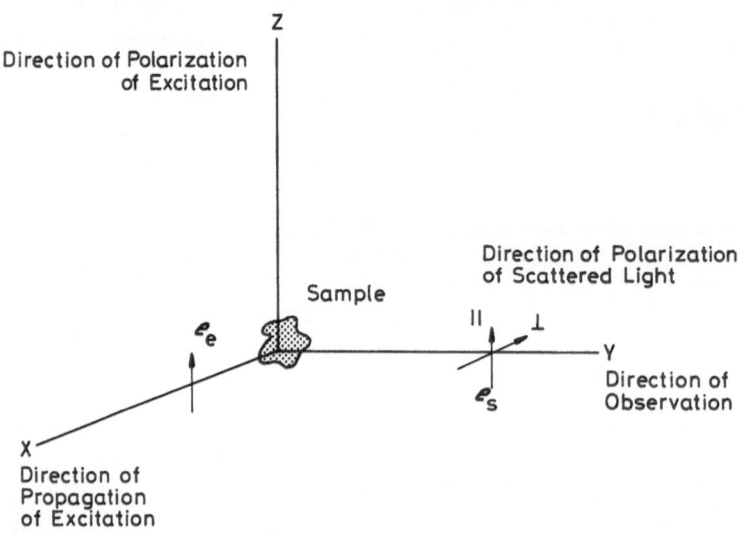

Fig. 2.1. Traditional Raman-scattering geometry

terms can be neglected. Furthermore, if magnetic phenomena are not involved, the polarizability tensor is composed of nine real elements [2]. The form of the tensor depends on the coordinate system chosen.

In the traditional Raman experiment the scattered light is observed in a direction perpendicular to that of the excitation. Furthermore, since laser sources are usually linearly polarized, the geometry shown in Fig. 2.1 represents a typical experiment. For this configuration Eq. (1) yields a Raman intensity which is proportional to

$$K\,|\tilde{e}_e \alpha e_s|^2 = K \left| (0 \ \ 0 \ \ 1) \begin{pmatrix} \alpha_{XX} & \alpha_{XY} & \alpha_{XZ} \\ \alpha_{YX} & \alpha_{YY} & \alpha_{YZ} \\ \alpha_{ZX} & \alpha_{ZY} & \alpha_{ZZ} \end{pmatrix} \begin{pmatrix} 1 \\ 0 \\ 1 \end{pmatrix} \right|^2 = K\,|\alpha_{ZX} + \alpha_{ZZ}|^2 .$$

(3)

Thus, only the two elements α_{ZX} and α_{ZZ} of the polarizability tensor contribute to light scattering in the Y direction. Finally, with the addition of an analyzer to select the direction of polarization of the scattered light, a particular element of α can be isolated. For example, if the analyzer is set in the Z direction, the intensity of the scattered light will depend only on the value of α_{ZZ}. Similarly, an analyzer setting in the X direction will yield a scattered intensity which depends only on α_{ZX}. These two cases correspond to measurements of the scattered intensities $\mathscr{I}_{\parallel}$ or $\mathscr{I}_{\perp}$ in which the electric vector of the scattered light is respectively parallel or perpendicular to the polarization direction of the excitation.

The coordinates X, Y and Z employed above define a laboratory system of axes. However, it is often convenient to specify the polarizability tensor with respect to axes x, y and z attached to a molecule in the sample; or, in the case of single-crystal samples the crystallographic axes may be used as a basis. The transformation of the polarizability tensor from one coordinate system to another can be carried out with the aid of the relation

$$\alpha_{xyz} = \Phi \alpha \tilde{\Phi} \,, \tag{4}$$

where following Wilson, Decius and Cross [3]

$$\Phi = \begin{pmatrix} \cos\theta\cos\varphi\cos\chi - \sin\varphi\sin\chi & \cos\theta\sin\varphi\cos\chi + \cos\varphi\sin\chi & -\sin\theta\cos\chi \\ -\cos\theta\cos\varphi\sin\chi - \sin\varphi\cos\chi & -\cos\theta\sin\varphi\sin\chi + \cos\varphi\cos\chi & \sin\theta\sin\chi \\ \sin\theta\cos\varphi & \sin\theta\sin\varphi & \cos\theta \end{pmatrix}$$

is an orthogonal matrix whose elements are direction cosines expressed in terms of Euler's angles (see Fig. 2.2).

A geometrical description of the polarizability tensor is often employed because of the analogy between α_{xyz} and the moment-of-inertia tensor in classical mechanics. The choice of the so-called principal axes of the ellipsoid is in each case related to the symmetry of the body considered (here, a molecule or crystal). An axis of three-fold or higher symmetry is necessarily a principal axis of the body. When principal axes are used, α_{xyz} is reduced to diagonal form. Furthermore, two special cases can be defined. If two of the diagonal elements are equal, e.g., $\alpha_{xx} = \alpha_{yy} \neq \alpha_{zz}$, the ellipsoid

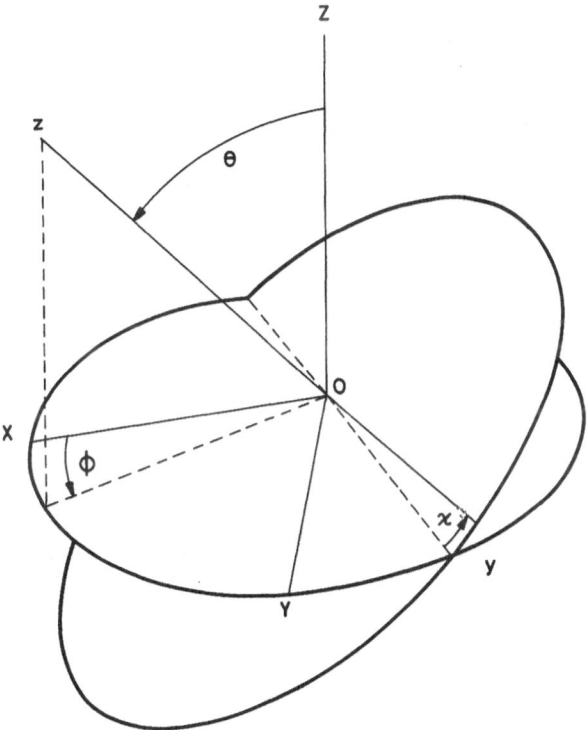

Fig. 2.2. Euler's angles

is a surface of revolution about the z axis. When $\alpha_{xx} = \alpha_{yy} = \alpha_{zz}$, the ellipsoid degenerates into a sphere and the polarizability of the molecule is characterized by a scalar quantity.

In the classical theory of the vibrational Raman effect, it is the derivative of the polarizability with respect to normal coordinates which is responsible for the scattering. Thus, the relative intensity of a given Raman band is determined by the square of quantities $\alpha'_{ij} \equiv \left(\dfrac{\partial \alpha_{ij}}{\partial q_v}\right)_0$, where q_v is the corresponding normal coordinate. In the quantum-mechanical treatment of the problem the analogous vibrational scattering (or "Raman") tensor can be derived. It should be noted that the symmetry properties of the polarizability tensor and the scattering tensor are identical, a consideration which is important in the determination of the selction rules for Raman activity. In the following section the analysis applies to a given Raman fundamental and the matrix of α''_{ij}'s will be represented simply by α.

2.2 Gaseous and Liquid Samples

If it can be assumed that the molecules in the sample are randomly oriented, as in low-pressure gases and, usually, even in liquids, Eq. (4) can be averaged over the Euler angles. In this case the individual elements of α_{xyz} cannot be evaluated experimentally. However, certain combinations of them which are invariant under coordinate transformations can be determined. In the notation of Mortensen and Hassing [4] the tensor invariants are defined by

$$\Sigma^0 = \frac{1}{3} |\alpha_{xx} + \alpha_{yy} + \alpha_{zz}|^2 , \tag{6}$$

$$\Sigma^1 = \frac{1}{2} \{|\alpha_{xy} - \alpha_{yx}|^2 + |\alpha_{yz} - \alpha_{zy}|^2 + |\alpha_{zx} - \alpha_{xz}|^2\} \tag{7}$$

and

$$\Sigma^2 = \frac{1}{2} \{|\alpha_{xy} + \alpha_{yx}|^2 + |\alpha_{yz} + \alpha_{zy}|^2 + |\alpha_{zx} + \alpha_{xz}|^2\}$$
$$+ \frac{1}{3} \{|\alpha_{xx} - \alpha_{yy}|^2 + |\alpha_{yy} - \alpha_{zz}|^2 + |\alpha_{zz} - \alpha_{xx}|^2\} . \tag{8}$$

It should be noted that the quantities Σ^k defined here are proportional to the invariants originally employed by Placzek [1]. They are related to other often-used parameters [2] by $\Sigma^0 = 3\bar{\alpha}^2$, $\Sigma^1 = \frac{2}{3}\delta^2$ and $\Sigma^2 = \frac{2}{3}\gamma^2$. The isotropic invariant $\bar{\alpha}$ is given by Eq. (12) of Chapter 1, while the anisotropic combination $\gamma(\alpha)$ defined there by Eq. (13) becomes equal to γ in the special case in which α is symmetric. In this chapter the quantity Σ^0 is used to represent the isotropic part of the tensor, while Σ^2 is the more general form of the symmetric part of the anisotropy. These invariants enter in the intensity formulae for normal Raman spectroscopy. The antisymmetric part of the tensor, which is represented here by Σ^1, plays a key role in both the resonance-Raman and the electronic-Raman effects, although it is equal to zero in "ordinary" Raman spectroscopy.

As the scattering elements enter to the second degree in the Raman intensity [Eq. (1)], the needed averages over molecular orientation take the form

$$\overline{\bar{\alpha}_{ij}^2} = \sum_k C_{ij}^k \Sigma^k, \qquad k = 0, 1, 2. \tag{9}$$

Although the coefficients C_{ij}^k can be determined by direct calculation with the aid of Euler's angles, they are more easily evaluated from the Clebsch-Gordon coefficients [5]. The results are given in Table 2.1.

Table 2.1 The coefficients C_{ij}^k in Eq. (9)

Conditions on i, j	$k = 0$	$k = 1$	$k = 2$
$i = j$	1/3	0	2/15
$i \neq j$	0	1/6	1/10

For the perpendicular scattering geometry defined in Fig. 2.1 the measured scattered intensities become

$$\mathcal{I}_{\parallel} \propto K \overline{\alpha_{zz}^2} = K \left(\frac{1}{3} \Sigma^0 + \frac{2}{15} \Sigma^2 \right) \tag{10}$$

and

$$\mathcal{I}_{\perp} \propto K \overline{\alpha_{zx}^2} = K \left(\frac{1}{6} \Sigma^1 + \frac{1}{10} \Sigma^2 \right), \tag{11}$$

where K is given below Eq. (1). As $\Sigma^1 = 0$ in ordinary (nonresonant) Raman scattering, a measurement of $\mathcal{I}_{\perp}$ yields directly a relative value of the anisotropy. Thus,

$$\mathcal{I}_{\perp} = \mathcal{I}_{\text{aniso.}} \propto \frac{K}{10} \Sigma^2 \tag{12}$$

and

$$\mathcal{I}_{\parallel} = \mathcal{I}_{\text{iso.}} + \frac{4}{3} \mathcal{I}_{\text{aniso.}}, \tag{13}$$

where $\mathcal{I}_{\text{iso.}} \propto \frac{K}{3} \Sigma^0$. It is often useful to evaluate $\mathcal{I}_{\text{iso.}}$ and $\mathcal{I}_{\text{aniso.}}$ directly from spectra recorded under the two different polarizing conditions [6]. From the above definitions

$$\mathcal{I}_{\text{iso.}} = \mathcal{I}_{\parallel} - \frac{4}{3} \mathcal{I}_{\perp} \tag{14}$$

and

$$\mathcal{I}_{\text{aniso.}} = \mathcal{I}_{\perp}. \tag{15}$$

In terms of the tensor invariants introduced above the traditional depolarization ratio, which is defined by

$$\varrho = \frac{\mathscr{I}_\perp}{\mathscr{I}_{||}}, \tag{16}$$

becomes

$$\varrho = \frac{5\Sigma^1 + 3\Sigma^2}{10\Sigma^0 + 4\Sigma^2} \tag{17}$$

for the case in which the molecules are randomly oriented in the scattering medium. It is apparent that in resonance-Raman spectroscopy ϱ can take on any positive value. However, in the ordinary Raman effect $\Sigma^1 = 0$ and $0 \leq \varrho \leq \frac{3}{4}$ for linearly polarized excitation. Furthermore, for Raman bands arising from non-totally symmetric molecular vibrations $\Sigma^0 = 0$, yielding $\varrho = 3/4$ for these so-called depolarized bands.

In the present analysis the traditional 90° scattering geometry has been assumed. However, simple geometrical considerations, combined with the averaging over molecular orientations [Eq. (9)] show that:

1. The depolarization ratio defined by Eq. (16) is independent of the scattering angle if the incident radiation is polarized in a direction perpendicular to the scattering plane, and

2. The observed depolarization ratio for any general scattering geometry is a simple function of that given by Eq. (16). Thus, no new information is obtained by varying the scattering angle and the antisymmetric tensor invariant Σ^1 cannot be determined with the use of only plane-polarized light. The addition of a measurement with the aid of circularly polarized light will, however, permit Σ^1 to be evaluated (see Sect. 2.11.).

In general the bands which are observed in the Raman spectra of gases and liquids are broadened due to the finite lifetime of the intermediate state involved in the scattering process, as well as to molecular motions. The actual profile of a Raman band then contains information concerning the scattering molecules and their inter-actions with their neighbors [7]. However, the measurement and analysis of band shapes is very difficult and probably does not fall within the domain of "Practical Raman Spectroscopy". Nevertheless, the term "intensity", as used above, needs clarification. Strictly speaking, the intensity of a band is obtained by integrating its profile over the entire frequency spectrum. From a practical point of view this pro-cedure is only applicable to bands which are well separated from all others, as band overlap can introduce serious errors. For this reason, and for simplicity, depolariza-tion ratios are often estimated from experimental spectra with the use of peak heights, rather than absolute intensities. It should be emphasized that this procedure is only correct if the bands being compared are symmetrical about their (common) peak frequency and if they have the same widths at half height. It is not obvious that instru-mentation which is designed to record automatically the depolarization ratio of the scattered light as a function of frequency can yield accurate values of the depolariza-tion ratios of Raman bands. On the other hand this method is very useful in that the resulting approximate values of the depolarization ratios provide immediate aids in assigning vibrational Raman spectra.

The width Γ of the isotropic component of a Raman band [whose intensity is represented by $\mathscr{I}_{\text{iso.}}$ in Eq. (14)] is given by $\Gamma = (4\pi c\tau)^{-1}$, where τ is the lifetime of the intermediate state involved. Thus, the profile of this component is of theoretical interest and can be obtained experimentally in ordinary Raman spectroscopy with the use of Eq. (14). The anisotropic component contains the molecular dynamics information, viz., rotation and orientation, and will not in general have the same bandwidth. It is apparent, then, that the accurate measurement of depolarization rations requires considerable care.

2.3 Single-Crystal Samples [8]

At a point r in a transparent crystal the electric field of a plane monochromatic wave propagating in a direction $\boldsymbol{K}_e$ is given by

$$\mathscr{E} = \mathscr{E}^0 \, e^{-2\pi i(\boldsymbol{K}_e \cdot \boldsymbol{r} - v_e t)} . \tag{18}$$

This expression is obtained by generalizing Eq. (2) of Chapter 1 to include the spatial characteristics of the wave. The frequency v_e of this exciting radiation is usually chosen to be in the visible region of the spectrum. Thus, v_e is very large compared to the frequencies of any of the crystal vibrations. If it is assumed here that the crystal is transparent at the frequency of the excitation, its refractive index n_e is real. The magnitude of the wave vector $\boldsymbol{K}_e$ is then given by

$$K_e = \frac{n_e}{c} v_e \tag{19}$$

and c/n_e is the velocity of propagation of the wave.

The classical treatment of the Raman effect in crystals now follows the general outline presented in Chap. 1. The polarizability tensor is expanded in normal coordinates, which are now represented by

$$q_v = A_v \, e^{\pm 2\pi i(\boldsymbol{k}_v \cdot \boldsymbol{r} - v_e t)} , \tag{20}$$

where $\boldsymbol{k}_v$ is the wave vector of lattice wave v. The induced dipole moment is then given to first order by

$$\mu_i = \alpha_0 \mathscr{E}^0 \, e^{-2\pi i(\boldsymbol{K}_e \cdot \boldsymbol{r} - v_e t)} + \mathscr{E}^0 \sum_v \alpha_v' A_v \, e^{2\pi i[(\boldsymbol{K}_e \pm \boldsymbol{k}_v) \cdot \boldsymbol{r} - (v_e \mp v_v) t]} , \tag{21}$$

where $\alpha_v' \equiv \left(\dfrac{\partial \alpha}{\partial q_v} \right)_0$. This expression is analogous to Eq. (7) of Chapter 1. It shows

that in a crystal the scattered light is of frequency $v_e \mp v_v$ and that it propagates in the direction given by $\boldsymbol{K}_e \pm \boldsymbol{k}_v$. When v_v is an optical lattice frequency, the scattering

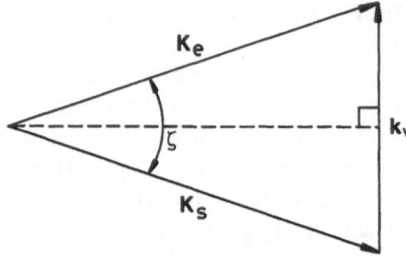

Fig. 2.3. Wave vectors for Raman scattering by a crystal [see Eq. (23)]

processes constitute the Raman effect. When acoustical frequencies are involved, the term Brillouin scattering is employed. As before, the frequencies given by $v_e \mp v_v$ are called Stokes or anti-Stokes frequencies, respectively, depending on the choice of sign.

The inelastic scattering processes in a crystal can be analyzed by considering the geometry shown in Fig. 2.3. If K_e and K_s are, respectively, the wave vectors of the exciting and scattered radiation, and k_v is the wave vector of phonon v, the conservation of energy and momentum requires that

$$v_e = v_s + v_v \tag{22}$$

and

$$K_e = K_s + k_v, \tag{23}$$

respectively. Since $v_v \ll v_e$, $K_e \simeq K_s$ and the magnitude of the wave vector of the scattering phonon is given by

$$k_v = 2K_e \sin \frac{\zeta}{2}, \tag{24}$$

where the angle ζ defines the scattering geometry. For the traditional 90° configuration Eq. (24) reduces to

$$k_v = \sqrt{2} \, K_e, \tag{25}$$

and the scattering wave in the crystal makes an angle of 135° with respect to the exciting beam. Thus, as lattice waves have directional properties, even in the case of cubic crystals, which are optically isotropic, the nature of the Raman spectrum depends on the orientation of the crystallographic axes with respect to the direction and polarization of the exciting radiation.

Some years ago Porto [9] proposed the notation which is in general used to specify the orientation of a single crystal in a given Raman experiment. Three directions in the crystal are defined — for example, x, y and z, which are not necessarily coincident with the crystallographic axes. For the geometry of Fig. 2.1, the excitation along the X axis is polarized in the Z direction, while the scattering is observed in the Y direction. The analyzer is set along either the X or the Z direction. If the crystal is oriented so

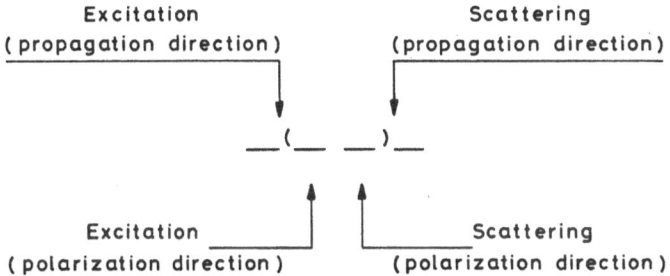

Fig. 2.4. Porto's notation [Ref. 9]

that the axes x, y, z are collinear with the space-fixed axes X, Y, Z, these two geo-metrical configurations are specified by x(zx)y and x(zz)y, respectively.

Clearly, the first symbol refers to the direction of propagation of the exciting radia-tion, while the last one refers to the observation direction. The symbols in parentheses specify, in order, the polarization directions of the exciting and scattered light (see Fig. 2.4.). Furthermore, they identify the particular element of the Raman tensor which is responsible for the observed scattering.

From the above discussion it should be apparent that Raman spectroscopic measure-ments on single crystals yield considerably more information than can be obtained from similar studies of liquids or gases. In the latter cases only two experimental configurations are possible, leading to the determination of the depolarization ratio. Single-crystal studies can, by suitable choices of the orientation, yield relative measure-ments of the squares of all of the elements of the scattering tensor.

A particularly useful table has been derived [10] which presents the form of the scattering tensors for each of the 32 crystal classes (symmetry point groups). It is reproduced here as Table 2.2, with appropriate corrections [11]. Each irreducible representation (symmetry species) in each class has a tensor of a specific form with respect to crystallographic axes. For example, for a tetragonal crystal of class 4 mmm $\equiv \mathcal{D}_{4h}$ the scattering tensors have the forms

$$A_{1g}: \begin{pmatrix} a & & \\ & a & \\ & & b \end{pmatrix}, \quad B_{1g}: \begin{pmatrix} c & & \\ & -c & \\ & & \end{pmatrix}, \quad B_{2g}: \begin{pmatrix} & d & \\ d & & \\ & & \end{pmatrix}$$

$$E_{g}: \begin{pmatrix} & & e \\ & & \\ e & & \end{pmatrix} \quad \text{and} \quad \begin{pmatrix} & & \\ & & e \\ & e & \end{pmatrix}.$$

Therefore, in the examples presented above the configuration x(zx)y will yield only vibrational fundamentals of species E_g, while the x(zz)y arrangement will result in activity of only the A_{1g}-species vibrations. To identify vibrations of other symmetry species in a crystal of this class, other orientations must be studied. For example, the configuration y(xy)z will result in a spectrum of the B_{2g}-species vibrations, and,

Table 2.2 Raman-active vibrational symmetries and Raman tensors for the crystal symmetry classes

System	Class		Raman tensors
Monoclinic	2	$\mathscr{C}_2$	$\begin{pmatrix} a & & d \\ & b & \\ d & & c \end{pmatrix}$ $\quad$ $\begin{pmatrix} & e & \\ e & & f \\ & f & \end{pmatrix}$
	m	$\mathscr{C}_s$	$A(Y)$ $\qquad$ $B(X,Z)$
	2/m	$\mathscr{C}_{2h}$	$A'(X,Z)$ $\qquad$ $A''(Y)$
			A_g $\qquad\qquad$ B_g
Orthorhombic	222	$\mathscr{D}_2$	$\begin{pmatrix} a & & \\ & b & \\ & & c \end{pmatrix}$ $\begin{pmatrix} & d & \\ d & & \\ & & \end{pmatrix}$ $\begin{pmatrix} & & e \\ & & \\ e & & \end{pmatrix}$ $\begin{pmatrix} & & \\ & & f \\ & f & \end{pmatrix}$
	mm2	$\mathscr{C}_{2v}$	$A \qquad B_1(Z) \qquad B_2(Y) \qquad B_3(X)$
	mmm	$\mathscr{D}_{2h}$	$A_1(Z) \quad A_2 \qquad B_1(X) \qquad B_2(Y)$
			$A_g \qquad B_{1g} \qquad\; B_{2g} \qquad\; B_{3g}$
Trigonal	3	$\mathscr{C}_3$	$\begin{pmatrix} a & & \\ & a & \\ & & b \end{pmatrix}$ $\begin{pmatrix} c & d & e \\ d & -c & f \\ e & f & \end{pmatrix}$ $\begin{pmatrix} d & -c & -f \\ -c & -d & e \\ -f & e & \end{pmatrix}$
	$\bar{3}$	$\mathscr{S}_6$	$A(Z) \qquad E(X) \qquad\qquad E(Y)$
			$A_g \qquad E_g \qquad\qquad\; E_g$
	32	$\mathscr{D}_3$	$\begin{pmatrix} a & & \\ & a & \\ & & b \end{pmatrix}$ $\begin{pmatrix} c & & \\ & -c & d \\ & d & \end{pmatrix}$ $\begin{pmatrix} & -c & -d \\ -c & & \\ -d & & \end{pmatrix}$
	3m	$\mathscr{C}_{3v}$	$A_1 \qquad E(X) \qquad\qquad E(Y)$
	$\bar{3}$m	$\mathscr{D}_{3d}$	$A_1(Z) \quad E(Y) \qquad\qquad E(-X)$
			$A_{1g} \qquad E_g \qquad\qquad\; E_g$

Tetragonal

$4,\ \bar{4},\ 4/m$ ($\mathscr{C}_4,\ \mathscr{S}_4,\ \mathscr{C}_{4h}$)

$$\begin{pmatrix} a & & \\ & a & \\ & & b \end{pmatrix}\quad\begin{pmatrix} c & d & \\ d & -c & \\ & & \end{pmatrix}\quad\begin{pmatrix} & & e \\ & & f \\ -f & -e & \end{pmatrix}\begin{pmatrix} & & -f \\ & & e \\ e & f & \end{pmatrix}$$

	$A(Z)$	B	$E(X)$	$E(Y)$
	A	$B(Z)$	$E(X)$	$E(-Y)$
	A_g	B_g	E_g	E_g

$4mm,\ 422,\ \bar{4}2m,\ 4/mmm$ ($\mathscr{C}_{4v},\ \mathscr{D}_4,\ \mathscr{D}_{2d},\ \mathscr{D}_{4h}$)

$$\begin{pmatrix} a & & \\ & a & \\ & & b \end{pmatrix}\quad\begin{pmatrix} c & & \\ & -c & \\ & & \end{pmatrix}\quad\begin{pmatrix} & d & \\ d & & \\ & & \end{pmatrix}\quad\begin{pmatrix} & & e \\ & & \\ e & & \end{pmatrix}\begin{pmatrix} & & \\ & & e \\ & e & \end{pmatrix}$$

	$A_1(Z)$	B_1	B_2	$E(X)$	$E(Y)$
	A_1	B_1	B_2	$E(-Y)$	$E(X)$
	A_1	B_1	$B_2(Z)$	$E(Y)$	$E(X)$
	A_{1g}	B_{1g}	B_{2g}	E_g	E_g

Hexagonal

$6,\ \bar{6},\ 6/m$ ($\mathscr{C}_6,\ \mathscr{C}_{3h},\ \mathscr{C}_{6h}$)

$$\begin{pmatrix} a & & \\ & a & \\ & & b \end{pmatrix}\quad\begin{pmatrix} c & d & \\ d & -c & \\ & & \end{pmatrix}\quad\begin{pmatrix} & & -d \\ & & c \\ -d & c & \end{pmatrix}\begin{pmatrix} & & e \\ & & f \\ f & -e & \end{pmatrix}$$

	$A(Z)$	$E_1(X)$	$E_1(Y)$	E_2	E_2
	A'	E''	E''	E_2	E_2
	A_g	E_{1g}	E_{1g}	E_{2g}	E_{2g}

$622,\ 6mm,\ \bar{6}m2,\ 6/mmm$ ($\mathscr{D}_6,\ \mathscr{C}_{6v},\ \mathscr{D}_{3h},\ \mathscr{D}_{6h}$)

$$\begin{pmatrix} a & & \\ & a & \\ & & b \end{pmatrix}\quad\begin{pmatrix} & & c \\ & & \\ c & & \end{pmatrix}\begin{pmatrix} & & \\ & & c \\ & c & \end{pmatrix}\quad\begin{pmatrix} d & & \\ & -d & \\ & & \end{pmatrix}\begin{pmatrix} & d & \\ d & & \\ & & \end{pmatrix}$$

	A_1	$E_1(X)$	$E_1(Y)$	E_2	E_2
	$A_1(Z)$	$E_1(Y)$	$E_1(-X)$	E_2	E_2
	A_1'	E''	E''	E_2'	$E'(Y)$
	A_{1g}	E_{1g}	E_{1g}	E_{2g}	E_{2g}

(Continued)

Table 2.2 (continued)

System	Class	Raman tensors

Cubic

$$\begin{pmatrix} a & & \\ & a & \\ & & a \end{pmatrix} \quad \begin{pmatrix} b & & \\ & b & \\ & & -2b \end{pmatrix} \begin{pmatrix} -\sqrt{3}b & & \\ & \sqrt{3}b & \\ & & 0 \end{pmatrix} \begin{pmatrix} & & \\ & & d \\ & d & \end{pmatrix} \begin{pmatrix} & & d \\ & & \\ d & & \end{pmatrix} \begin{pmatrix} & d & \\ d & & \\ & & \end{pmatrix}$$

Class		Symmetry species
23	$\mathcal{T}$	A E E F(X) F(Y) F(Z)
m3	$\mathcal{T}_h$	A_g E_g E_g F_g F_g F_g
432	$\mathcal{O}$	A_1 E E F_2 F_2 F_2
$\bar{4}3m$	$\mathcal{T}_d$	A_1 E E $F_2(X)$ $F_2(Y)$ $F_2(Z)$
m3m	$\mathcal{O}_h$	A_{1g} E_g E_g F_{2g} F_{2g} F_{2g}

finally, the y(xx)z arrangement will yield a spectrum consisting of both A_{1g} and B_{1g} species vibrations. This result, when combined with the spectrum obtained above from the x(zz)y orientation, allows the symmetry species of all vibrations to be determined. Obviously, this method provides a very powerful experimental tool for the assignment of the vibrational Raman bands in the spectra of a single crystal.

The polarized Raman spectra of MnF_2 are shown in Fig. 2.5. in order to illustrate the principle outlined above. The crystal class of this compound is $\mathscr{D}_{4h}$ and it is found by factor-group analysis [8] to have one Raman-active fundamental vibration of each of the four symmetry species. Thus, the use of polarization techniques in this case allows unambiguous assignments of the observed Raman bands to be made.

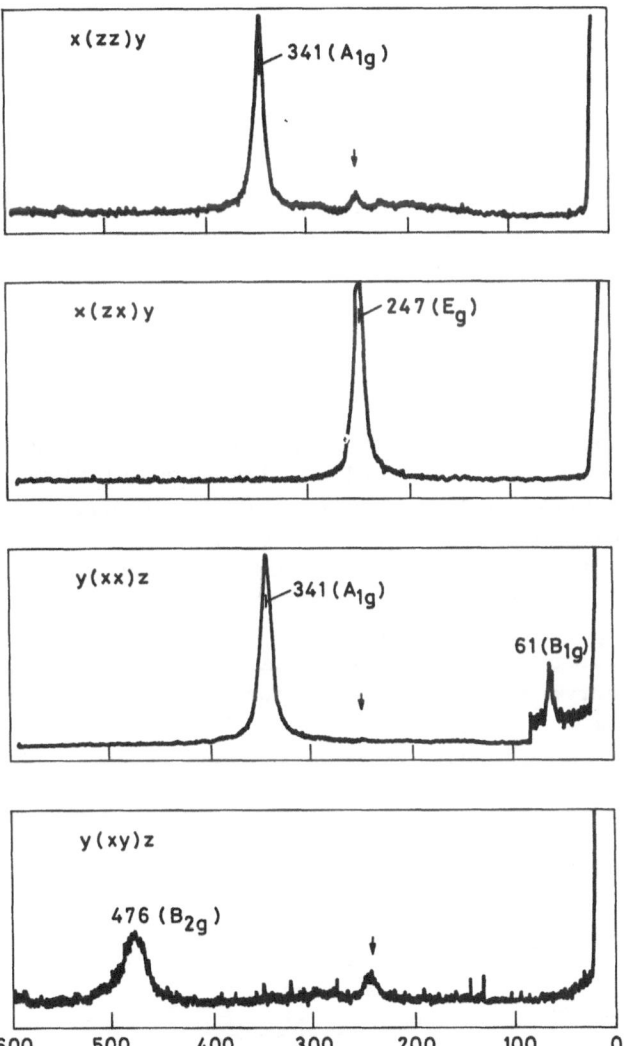

Fig. 2.5. Raman spectra of MnF_2 [From Ref. 12]

If the Raman fundamental being investigated is also infrared active, further considerations enter. In this case the dipole derivative $\left(\dfrac{\partial\mu}{\partial q_v}\right)_0$ is nonzero and the electric field of the incident electromagnetic radiation interacts with the crystal vibrations. The transverse and longitudinal components of the vibrations then have different frequencies, v_a and v_l, respectively, whose separation is proportional to $\left(\dfrac{\partial\mu}{\partial q_v}\right)_0^2$ and, hence, to the absorption intensity of the corresponding infrared band. In molecular crystals this effect is usually small. However, in ionic crystals, which often exhibit strong infrared absorption, the difference $v_l - v_t$, which corresponds approximately to the width of the Reststrahlen region, can become quite large.

It should be noted that the requirement of simultaneous infrared and Raman activity limits this phenomenon to piezoelectric crystals. Otherwise, for the ten centrosymmetric crystal classes the rule of mutual exclusion applies and no fundamental vibration can be both infrared and Raman active. In piezoelectric crystals, then, each Raman-active fundamental predicted by factor-group analysis leads to a pair of Raman lines (at frequencies v_l and v_t), while only the band at v_t appears in infrared absorption. As the analysis of Raman polarizations is quite complicated in this case, the reader is refered to the excellent review article by Loudon [10] for the details.

2.4 Excitation Focusing

Thus far in this chapter it has been assumed that Raman scattering from a given sample is excited by a parallel beam of linearly polarized light. The output of the rare-gas ion lasers usually used in Raman spectroscopy, when operated in the fundamental (TEM_{00}) mode, fulfills this condition to a very good approximation. However, inspection of almost any Raman spectrometer reveals that the laser beam is in fact focused on the sample with the use of a lens.

As the intensity of Raman scattering is proportional to the irradiance (power/area of cross section) of the exciting beam, it is usually advantageous to focus the beam on the sample [13, 14]. However, high irradiance may in some cases result in damage to the material being studied.

The "focal point" of the exciting beam is usually adjusted to coincide with the center of the sample. At this point the diameter of the beam is not zero, but is in reality

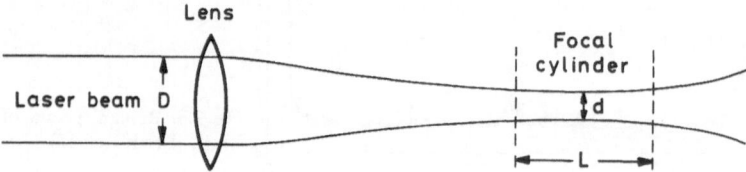

Fig. 2.6. Focusing of a laser beam

limited by aberation effects and, fundamentally, by the diffraction limit, which is a function of the wavelength of the light. It is, therefore, more realistic to consider the focal region to be approximately cylindrical, as indicated in Fig. 2.6. The diameter or "waist" of this focal cylinder is given by [15, 2]

$$d = \frac{4\lambda f}{\pi D},$$ (26)

where λ is the wavelength of the light, f is the focal length of the lens and D is the diameter of the laser beam. The effective length of the focal cylinder is given approximately by [15, 2]

$$L = \frac{16\lambda f^2}{\pi D^2}.$$ (27)

The lenses used to focus a laser beam generally have focal lengths of 10 to 30 cm. For a laser beam of 3 mm in diameter operating at a wavelength of 500 nm, Eq. (26) gives $d \simeq 40$ µm and the gain in irradiance, and hence, in Raman intensity, is by a factor of more than 10^3. From Eq. (27) the length of the focal cylinder is of the order of 1 cm and its volume in this typical case is approximately 10^{-5} cm^3.

As the focusing lenses used here do not have significant depolarizing effects, the beam can still be assumed to be linearly polarized within the focal cylinder. However, it should be noted that in the case of focusing by a very wide-aperture lens (such as a microscope objective) significant depolarization results. This problem is considered later in this chapter (Sect. 2.8).

2.5 Collection Optics

Equation (1) shows that the total Raman intensity depends on the solid angle Ω over which the scattered light is collected. For this reason a large, short-focal length lens is usually used to focus the light on the entrance slit of the monochromator. Accordingly, the term "direction of scattering", which was employed in the above discussion of polarization measurements, must be reconsidered. This problem has been treated by several authors [16–18], all of whom came to the conclusion that the error in a measured depolarization ratio due to the nonzero angle of collection is surprisingly small, although not necessarily negligible.

A simple analysis of the effects of the collection lens can be made with the assumption of a point source of Raman scattering in the center of the sample. This approximation is valid in most cases for determining the polarization properties of light scattered from nonabsorbing samples.

The direction of the electric vector associated with a ray of light within the cone of scattered light can be specified by the Euler angles defined in Fig. 2.2. Then,

$$e_s = \begin{pmatrix} -\cos\theta\cos\varphi\sin\chi - \sin\varphi\cos\chi \\ -\cos\theta\sin\varphi\sin\chi + \cos\varphi\cos\chi \\ \sin\theta\sin\chi \end{pmatrix}$$ (28)

and the intensity of the light scattered in the Y direction (see Fig. 2.1) is

$$\mathscr{I}_s \propto (e_s^X \alpha_{ZX} + e_s^Y \alpha_{ZY} + e_s^Z \alpha_{ZZ})^2 . \tag{29}$$

Integration over the cone yields [18]

$$\mathscr{I}_s \propto A\alpha_{ZX}^2 + B\alpha_{ZY}^2 + A\alpha_{ZZ}^2 , \tag{30}$$

where

$$A = \pi^2 \int_0^{\theta_m} (\cos^2 \theta + 1) \sin \theta \, d\theta = \pi^2 \left(\frac{4}{3} - \cos \theta_m - \frac{1}{3} \cos^3 \theta_m \right) \tag{31}$$

and*

$$B = 2\pi^2 \int_0^{\theta_m} \sin^3 \theta \, d\theta = 2\pi^2 \left(\frac{2}{3} - \cos \theta_m + \frac{1}{3} \cos^3 \theta_m \right) . \tag{32}$$

The upper limit appearing in the above integrals is the half angle of the scattered light cone, which is defined for a given lens by $\sin \theta_m = NA$ (numerical aperture) or $f/number = (2 \tan \theta_m)^{-1}$. With the aid of Eq. (9) and the coefficients given in Table 1, the average over molecular orientations is

$$\mathscr{I}_s \propto A \left(\frac{1}{6} \Sigma^1 + \frac{1}{10} \Sigma^2 \right) + B \left(\frac{1}{6} \Sigma^1 + \frac{1}{10} \Sigma^2 \right) + A \left(\frac{1}{3} \Sigma^0 + \frac{2}{15} \Sigma^2 \right) . \tag{33}$$

Thus, for gaseous and liquid samples the depolarization ratio becomes

$$\varrho' = \frac{(A + B)(5\Sigma^1 + 3\Sigma^2)}{10A\Sigma^0 + 5B\Sigma^1 + (4A + 3B)\Sigma^2} , \tag{34}$$

where it is assumed that the analyzer selects only the light polarized in either the Y, Z(∥) or the X, Y(⊥) planes. The parameter B in Eq. (34) arises from the contributions made by the component of the scattered light which is polarized in the "direction of observation" (Y). This result is to be compared with Eq. (17), which was obtained on the basis of the unrealistic assumption that the scattered light forms a parallel beam in the Y direction.

As an example of the application of these results, consider the normal Raman effect ($\Sigma^1 = 0$), for which

$$\varrho = \frac{3\Sigma^2}{10\Sigma^0 + 4\Sigma^2} \tag{35}$$

---

* Note that a factor of 2π was omitted in Eq. (13) of Ref. 18.

and

$$\varrho' = \frac{3\Sigma^2 (A + B)}{10A\Sigma^0 + (4A + 3B) \Sigma^2} .$$

(36)

The two limiting cases are
1. Strongly polarized bands ($10\Sigma^0 \gg 4\Sigma^2$), for which

$$\varrho' = \varrho \left(1 + \frac{B}{A} \right),$$

(37)

and
2. Depolarized bands ($\Sigma^0 = 0$), for which

$$\varrho' = \varrho \left(1 + \frac{B}{4A} + ... \right),$$

(38)

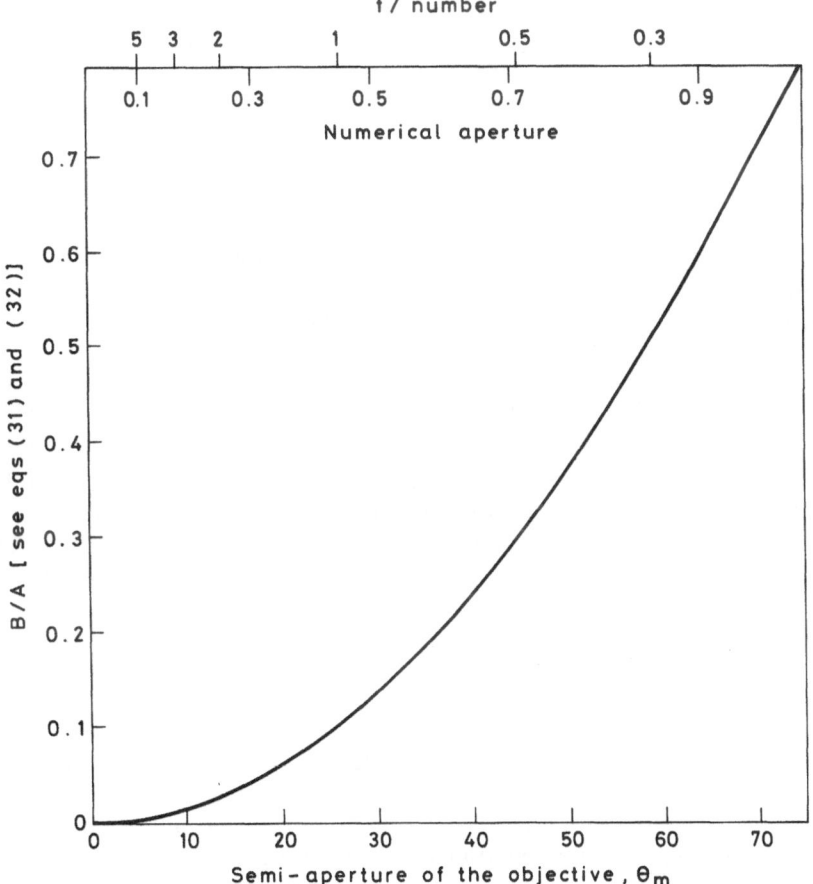

Fig. 2.7. Correction B/A vs semi-aperture of collecting lens, θ_m

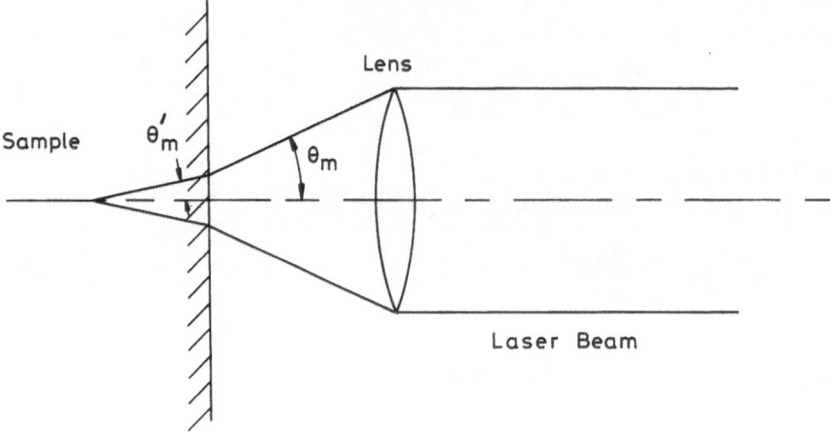

Fig. 2.8. Effective semi-aperture, θ'_m, for a sample of refractive index n

with $\varrho = 3/4$. It can be concluded, then, that the error due to a nonzero collection angle is more important for polarized bands and that it is, in general, a function of the true depolarization ratio of the band.

The variation of the ratio B/A with the half angle of the scattered cone is shown in Fig. 2.7. It is apparent that errors in polarization measurements which are due to the effect of the collecting lens become negligible for $\theta \lesssim 10°$, that is, for lenses with f/numbers $\gtrsim 3$.

For liquid samples with refractive indices much greater than unity the half angle of the cone, θ_m, is modified as illustrated in Fig. 2.8. The effective half angle can then be defined by

$$\theta'_m = \sin^{-1}\left(\frac{1}{n}\sin\theta_m\right), \tag{39}$$

where n is the (real) refractive index of the sample. The upper limits on the integrals defined by Eqs. (31) and (32) are then replaced by Eq. (39). It should be noted that the effect of an increased index of refraction of the sample is to reduce the errors in polarization measurements for a given collecting lens. This result is important in Raman microspectroscopy, as discussed in Sect. 2.8.

In the case of optically isotropic solid samples, or any sample whose Raman scattering is observed along a three-fold or higher symmetry axis, Eqs. (30–32) still apply. They provide a basis for the analysis of polarization measurements under various experimental configurations when a wide-aperture collection lens is employed. For the example considered in Sect. 2.3, the possible experiments are summarized in Table 2.3. The relative intensities, as determined from Table 2.2, are listed in the second column. The terms in B arise in each case from the nonparallel nature of the observed scattered light and thus contribute to the polarization leakage.

The spectra presented in Fig. 2.5 were obtained under the four experimental conditions shown in the first four lines of Table 2.3. In these cases the strong band asso-

Table 2.3 The "Porto experiments' for a crystal of class 4mmm ≡ $\mathscr{D}_{4h}$

Experiment	Relative Raman intensity	Principal species	Leakage species
x(zx) y	$(A + B)\, e^2$	E_g	E_g
x(zz) y	$Ab^2 + Be^2$	A_{1g}	E_g
y(xy) z	$Ad^2 + Be^2$	B_{2g}	E_g
y(xx) z	$A(a^2 + c^2) + Be^2$	$A_{1g} + B_{1g}$	E_g
z(yz) x	$Ae^2 + Bd^2$	E_g	B_{2g}
z(yy) x	$A(a^2 + c^2) + Bd^2$	$A_{1g} + B_{1g}$	B_{2g}
x(yx) y	$Ad^2 + B(a^2 + c^2)$	B_{2g}	$A_{1g} + B_{1g}$
x(yz) y	$Ae^2 + B(a^2 + c^2)$	E_g	$A_{1g} + B_{1g}$
y(zy) z	$Ae^2 + Bb^2$	E_g	A_{1g}
y(zx) z	$Ae^2 + Bb^2$	E_g	A_{1g}
z(xz) x	$Ae^2 + B(a^2 + c^2)$	E_g	$A_{1g} + B_{1g}$
z(xy) x	$Ad^2 + B(a^2 + c^2)$	B_{2g}	$A_{1g} + B_{1g}$

ciated with the E_g-species vibration appears in each of the spectra, as predicted from this analysis (see arrow).

2.6 Absorbing Samples

In the analysis of sample illumination which was presented above, it was assumed that neither the exciting light nor the scattered light was absorbed by the sample. For studies of such transparent samples the choice of scattering geometry is not of primary

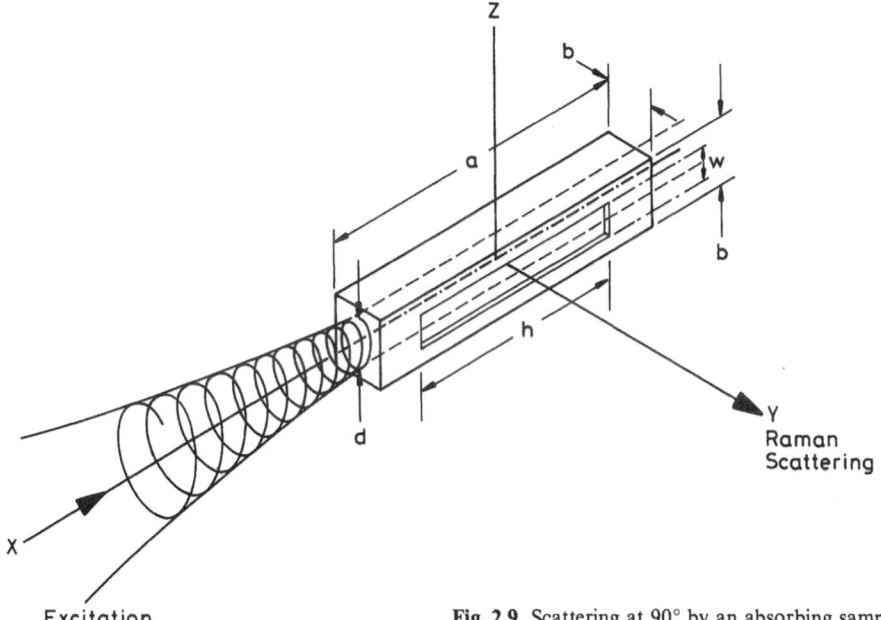

Excitation

Fig. 2.9. Scattering at 90° by an absorbing sample

importance. However, in the case of samples which absorb light at the excitation and/or scattering wavelengths, the experimental configuration must be carefully selected. Two basic problems arise: First, if absorbing solutions are being investigated, optimum concentrations should be chosen. Secondly, in the case of resonance of the excitation with electronic transitions (see Chap. 1.), polarization considerations are very different and must be analyzed accordingly, as shown in Sect. 2.11.

A simple analysis will now be made of the problem of light absorption by a Raman sample. The geometry will be taken to be the classical 90° scattering, as shown in Fig. 2.9. The sample, a liquid solution of concentration c moles/l, is contained in a cell of length a with square cross-section b × b. As shown below, the analysis applies equally to a cylindrical cell of length a and diameter b.

The laser beam is focused by a lens to a diameter d, which is defined by the 1/e-power condition on a Gaussian beam. As indicated above, under typical experimental conditions $d \simeq 0.04$ mm and, usually, $d < b$. The length of the sample cell, a, is chosen to be smaller than the length of the focal cylinder, viz., $a < L \simeq 1$ cm.

With the origin at the center of the sample the irradiance of the exciting beam at a point along the X axis is given by

$$I_e = I_e^0 \, e^{-c\varepsilon_e \left(\frac{a}{2} - x\right)}, \tag{40}$$

where c is the concentration of the solution in moles per liter and ε_e is the molar extinction coefficient of the solution at the wavelength of the excitation. If it is further assumed that the cross section of the laser beam has a Gaussian energy distribution, Eq. (40) becomes

$$I_e = I_e^0 \, e^{-c\varepsilon_e \left(\frac{a}{2} - x\right)} e^{-4(Y^2 + Z^2)/d^2}. \tag{41}$$

The intensity of the Raman-scattering by a volume dX dY dZ within the sample is given by

$$d\mathscr{I}_s^0 \propto I_e c \, dX \, dY \, dZ \tag{42}$$

and at a given point along the Y axis by

$$d\mathscr{I} = d\mathscr{I}_s^0 \, e^{-c\varepsilon_s \left(\frac{b}{2} - Y\right)}. \tag{43}$$

The image of the entrance slit of the spectrometer is defined by h and w in Fig. 2.9. Here, it is assumed that $w \ll d$, hence, the integration over dZ yields a factor w and Eqs. (41–43) become

$$d\mathscr{I}_s \propto cwI_e^0 \, e^{-c\varepsilon_s \left(\frac{b}{2} - Y\right) - 4Y^2/d^2} \, e^{-c\varepsilon_e \left(\frac{a}{2} - x\right)} dX \, dY. \tag{44}$$

The integration of this expression can be carried out directly in the form

$$\mathcal{I}_s \propto cwI_e^0 \, e^{-\frac{c}{2}(\varepsilon_e a + \varepsilon_s b)} \int_{-\frac{d}{2}}^{\frac{d}{2}} e^{-Y^2/d^2 + c\varepsilon_s Y} \, dY \int_{-\frac{h}{2}}^{\frac{h}{2}} e^{c\varepsilon_e X} \, dX \tag{45}$$

$$= \frac{wd}{\varepsilon_e} I_e^0 \, e^{-\frac{c}{2}(\varepsilon_e a + \varepsilon_s b)} \sinh (c\varepsilon_e h/2) \, e^{c^2 \varepsilon_s^2 d^2/16}$$

$$\times \frac{|\sqrt{\pi}}{2} \{ \mathrm{erf}\,(1 - c\varepsilon_e d/4) + \mathrm{erf}\,(1 + c\varepsilon_e d/4) \} \tag{46}$$

$$\simeq \frac{wd}{\varepsilon_e} I_e^0 \, e^{-\frac{c}{2}(\varepsilon_e a + \varepsilon_s b)} \sinh (c\varepsilon_e h/2) \tag{47}$$

for $c\varepsilon_s d \ll 4$. This condition is certainly a reasonable one in most cases.

The result of the above analysis, as expressed by Eq. (47), shows that at low concentrations the Raman intensity increases with concentration, but that it reaches a maximum at a certain value, c_m. This optimal concentration can be easily determined by setting the derivative of Eq. (47) with respect to c equal to zero. The result is

$$c_m = \frac{2}{\varepsilon_e h} \tanh^{-1} \left(\frac{\varepsilon_e h}{a\varepsilon_e + b\varepsilon_s} \right). \tag{48}$$

The simpler expression

$$c_m = \frac{2}{a\varepsilon_e + b\varepsilon_s} \tag{49}$$

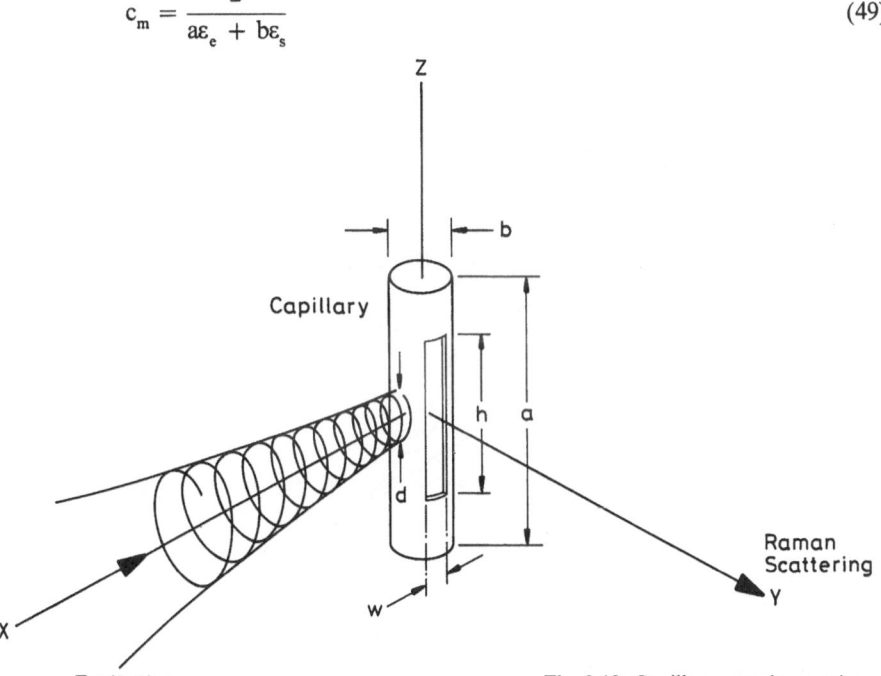

Fig. 2.10. Capillary sample container

given previously be several authors [19, 20] is obtained from Eq. (48) in the limit that $c\varepsilon_e h/2 \ll 1$.

A more conventional, but less efficient sample configuration is shown in Fig. 2.10. The solution is contained in a vertically mounted capillary and excitation and scattering are in the horizontal plane. With the circular cross section of the capillary Eq. (47) reduces to

$$\mathcal{I}_s \propto wd^2 I_e^0 c e^{-\frac{cb}{2}(\varepsilon_e + \varepsilon_s)} . \tag{50}$$

It should be noted that since $d < h$ in most cases, the slit is not completely filled and some loss in intensity occurs.

If the extinction coefficients at the exciting and scattering wavelengths do not differ greatly, the average extinction coefficient $\varepsilon = \dfrac{1}{2}(\varepsilon_e + \varepsilon_s)$ can be introduced and the scattered light intensity takes the form

$$\mathcal{I}_s \propto wd^2 I_e^0 c e^{-\varepsilon cb} . \tag{51}$$

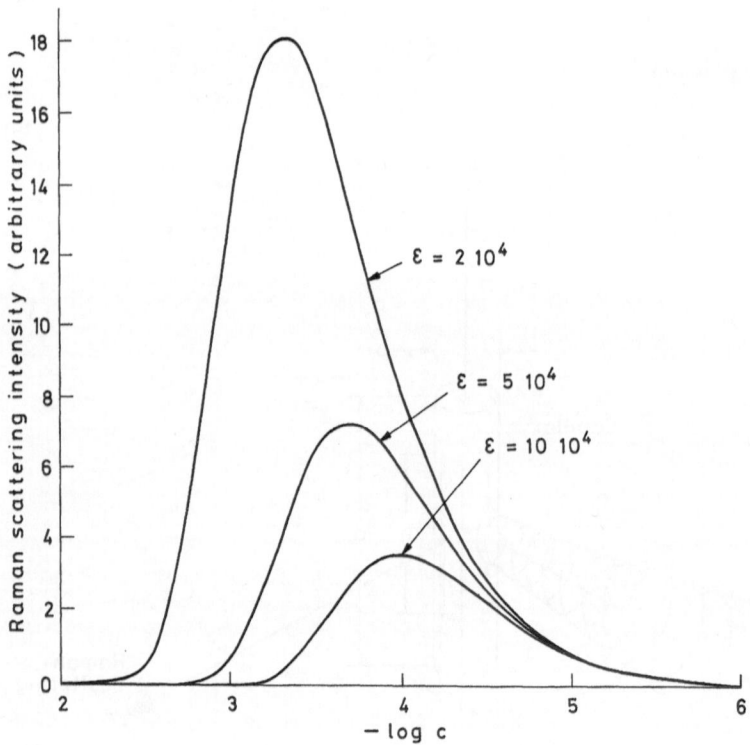

Fig. 2.11. Raman scattering vs concentration of absorbing solutes (90° configuration)

This function is plotted in Fig. 2.11 for typical values of ε and an assumed capillary diameter of 1 mm.

2.7 Backscattering by Absorbing Samples

In the examples considered thus far in this chapter, the classical 90° scattering geometry has been assumed. However, in certain applications other scattering angles can be advantageous. Raman spectra have been obtained with the use of the back-scattering technique (θ = 180°), as well as the seemingly unlikely forward scattering configuration (θ ≃ 0°). Furthermore, as shown in Sect. 2.9, with the exciting beam at glancing incidence with respect to a surface, valuable Raman spectroscopic information can be obtained from thin films and adsorbed species.

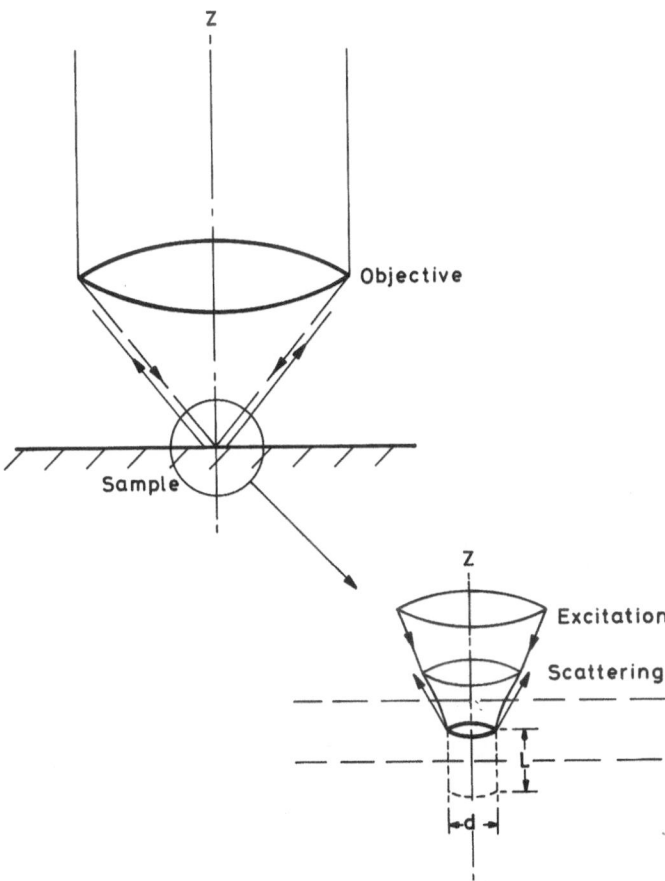

Fig. 2.12. Backscattering geometry

The backscattering geometry has proven to be especially useful in the study of highly absorbing samples. Furthermore, it is the basic configuration employed in micro-Raman instruments (see Chap. 6.). However, as shown in the following section, the wide-aperture objectives which are invariably used in Raman microspectrometers introduce other considerations. In this section the problem of absorbing samples in the backscattering configuration will be briefly analyzed.

Following the treatment of sample absorption presented above for the 90° configuration, the backscattering geometry will be considered here on the basis of Fig. 2.12. The laser excitation, which is assumed to consist of a parallel beam within the focal cylinder, is directed along the Z axis. The irradiance at a point within the sample is then given by

$$I_e = I_e^0 \, e^{-c\varepsilon_e Z} \, e^{-4(X^2+Y^2)/d^2} \,, \tag{52}$$

and the intensity of the Raman-scattered light by a volume dX dY dZ within the sample can be expressed by

$$d\mathscr{I}_s^0 \propto I_e c \, dX \, dY \, dZ \,. \tag{53}$$

The intensity of the scattered light leaving the sample in the Z direction is then

$$d\mathscr{I}_s = d\mathscr{I}_s^0 \, e^{-c\varepsilon_s Z} \,. \tag{54}$$

The combination of Eqs. (53) and (54) leads to the relation

$$d\mathscr{I} \propto I_e^0 c e^{-c(\varepsilon_s + \varepsilon_e)Z} \, dZ \, e^{-4(X^2+Y^2)/d^2} \, dX \, dY \,, \tag{55}$$

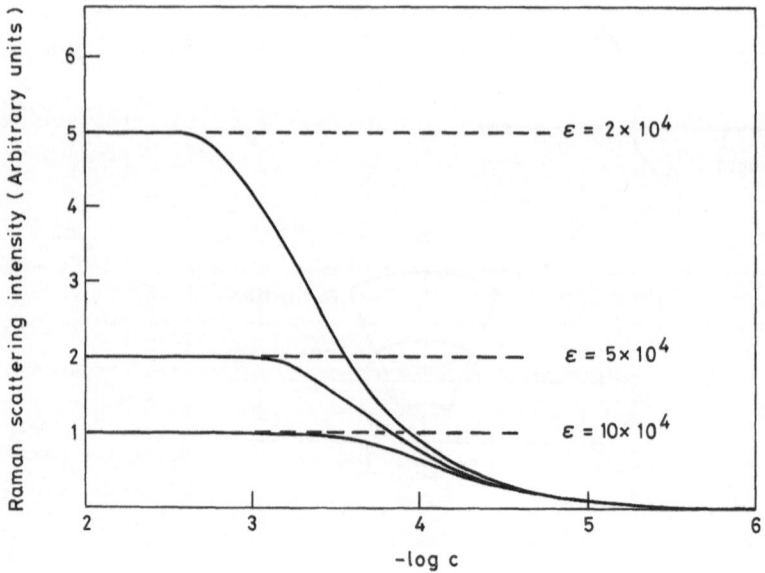

Fig. 2.13. Raman scattering vs concentration of absorbing solutes (180° configuration)

whose integration yields approximately

$$\mathcal{I}_\nu \propto \frac{\pi(1 - 1/e)}{16} \frac{I_e^0 d^2}{\varepsilon_e + \varepsilon_s} (1 - e^{-c(\varepsilon_e + \varepsilon_s)L}) ,$$ (56)

where L is here the effective depth of focus (see Fig. 2.12). This expression is represented in Fig. 2.13, where it is evident that for high concentrations of absorbing species, the backscattering geometry offers a distinct advantage over the usual 90° configuration (Fig. 2.11). This result was established experimentally many years ago by Hendra et al. [21].

2.8 Wide-Angle Objectives

The effect of focusing the exciting beam was considered briefly in Sect. 2.4. There it was pointed out that the use of a long-focal length lens to focus the radiation on the sample does not introduce significant depolarization. However, when wide-aperture lenses, such as microscope objectives are employed, corrections must be made for the resulting depolarization if Raman polarization measurements are to be made. This problem has been the subject of several recent publications [18, 22–24].

It has been shown that the electric vector of the excitation in the focal region within a sample is given by [25].

$$\mathscr{E} = \begin{pmatrix} -i(I_0 + I_2 \cos 2\psi) \\ -iI_2 \sin 2\psi \\ -2I_1 \cos \psi \end{pmatrix} ,$$ (57)

if the laser beam is plane polarized in the X direction before passing through the microscope objective. The integrals involved in Eq. (57) have been derived for the case of a nonabsorbing isotropic sample of refractive index n in the form [23]

$$I_0(u, v, n) = 2 \int_0^{\theta_m} D(\theta) \sin \theta \left[\frac{m}{n^2 \cos \theta + m} + \frac{1}{\cos \theta + m} \right]$$

$$\times \cos \theta \, J_0 \left[\frac{v \sin \theta}{\sin \theta_m} \right] e^{iu \cos \theta / \sin^2 \theta_m} \cos^{1/2} \theta \, d\theta .$$ (58)

$$I_1(u, v, n) = 2 \int_0^{\theta_m} D(\theta) \sin \theta \left[\frac{\sin \theta}{n^2 \cos \theta + m} \right] \cos \theta \, J_1 \left[\frac{v \sin \theta}{\sin \theta_m} \right]$$

$$\times e^{iu \cos \theta / \sin^2 \theta_m} \cos^{1/2} \theta \, d\theta$$ (59)

and

$$I_2(u, v, n) = \cdots 2 \int_0^{\theta_m} D(\theta) \sin\theta \left[\frac{m}{n^2 \cos\theta + m} - \frac{1}{\cos\theta + m} \right]$$

$$\times \cos\theta \, J_2 \left[\frac{v \sin\theta}{\sin\theta_m} \right] e^{iu \cos\theta / \sin^2\theta_m} \cos^{1/2}\theta \, d\theta . \tag{60}$$

where θ is the angle of incidence of a given ray of the light excitation, $D(\theta)$ $= N \csc\theta \exp(-\sin^2\theta / \sin^2\theta_m)$ represents a Gaussian radial distribution in the laser beam and N is a normalization constant. A point in the focal region is specified by the dimensionless cylindrical coordinates $u = kZ \sin^2\theta_m$, $v = k\sqrt{X^2 + Y^2} \sin\theta_m$ and angle ψ. Here $k = 2\pi/\lambda$ is the wavenumber, $m = \sqrt{n^2 - \sin\theta}$ and the J_n's are the Bessel functions of the first kind. The integrals defined by Eqs. (58–60) are functions of n, the (real, isotropic) refractive index of the sample.

The intensity of the Raman scattered light can be calculated from Eq. (1) with the aid of Eq. (28), which specifies the direction of the electric vector of the scattered radiation. In the backscattering configuration shown in Fig. 2.12, which is usually

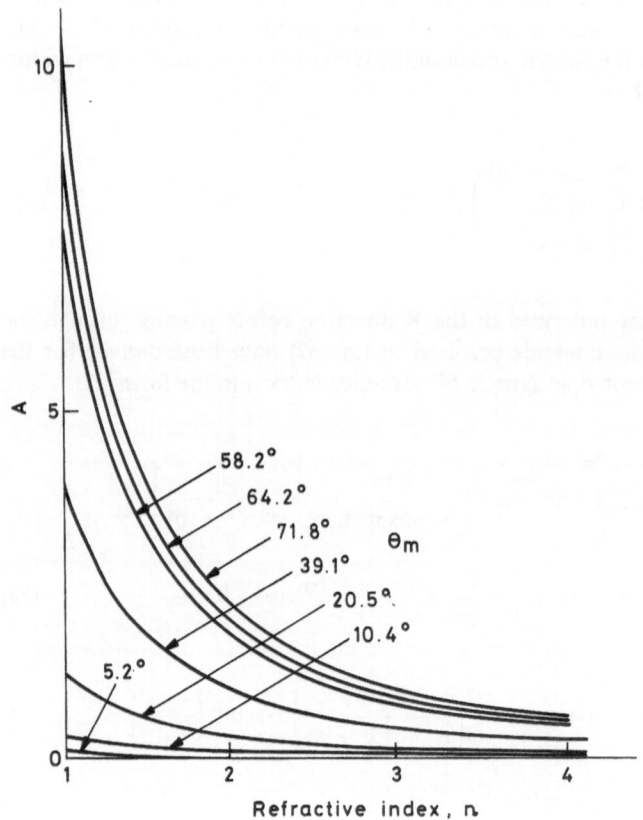

Fig. 2.14. Parameter A as a function of refractive index [See Eqs. (31) and (39).]

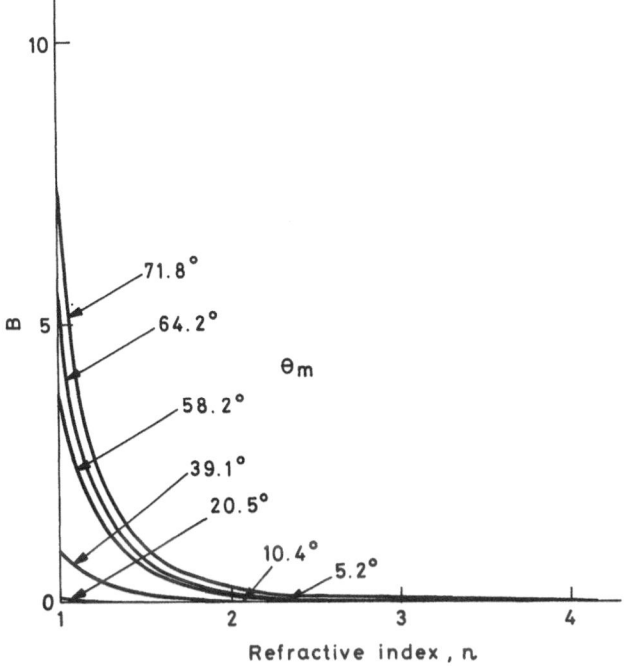

Fig. 2.15. Parameter B as a function of refractive index [See Eqs. (32) and (39).]

employed in micro-Raman instruments, the solid angle Ω describes approximately both the incident and scattered cones. If the cone axis is collinear with Z and the excitation is assumed to be polarized in the X direction, the expression for the relative Raman intensity becomes

$$
\begin{aligned}
\mathscr{I} = {} & (\alpha_{XX}^2 A + \alpha_{XY}^2 A + \alpha_{XZ}^2 B)\,(2C_0 + C_2) \\
& + (\alpha_{YX}^2 A + \alpha_{YY}^2 A + \alpha_{YZ}^2 B)\,C_2 \\
& + (\alpha_{ZX}^2 A + \alpha_{ZY}^2 A + \alpha_{ZZ}^2 B)\,4C_1
\end{aligned}
\tag{61}
$$

where

$$
C_j = 2\pi \int_0^\infty \int_0^\infty |\,I_j(u,\, v,\, n)|^2 \, v \, dv \, du\,, \qquad j = 0, 1, 2\,.
\tag{62}
$$

The parameters A and B are given by Eqs. (31) and (32), with θ_m replaced by θ_m' (see Eq. (39)), and plotted as functions of n in Figs. 2.14 and 2.15, respectively. The integrals C_j are plotted as functions of n in Figs. 2.16, 2.17 and 2.18.

Equation (61) serves as the basis for the interpretation of polarization measurements on isotropic media. If the scattered light is analyzed in the direction parallel

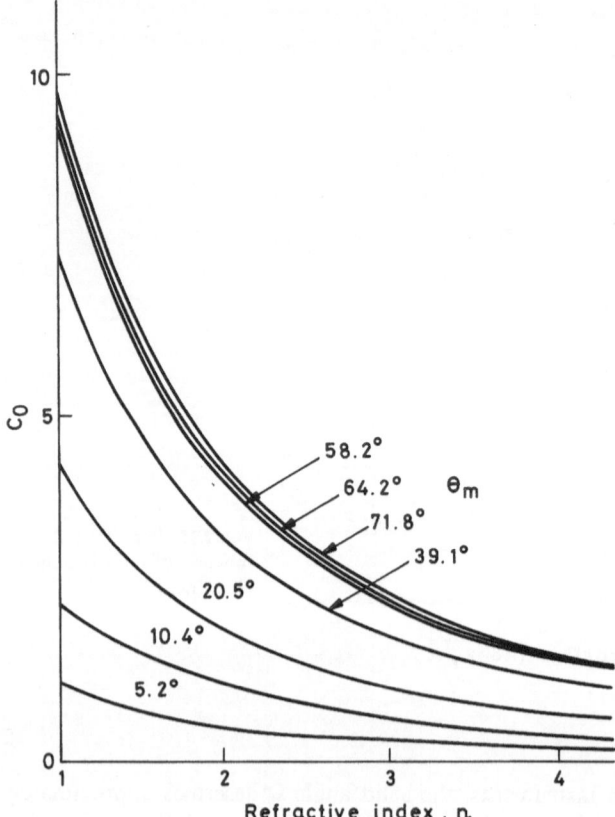

Fig. 2.16. Parameter C_0 as a function of refractive index [See Eq. (62).]

to the direction of polarization of the excitation (X axis), the scattered intensity is given by

$$\mathscr{I}_{\parallel} = (\alpha^2_{XX}A + \alpha^2_{XZ}B)(2C_0 + C_2) + (\alpha^2_{YX}A + \alpha^2_{YZ}B)C_2$$

$$+ (\alpha^2_{ZX}A + \alpha^2_{ZZ}B)4C_1 .$$

(63)

On the other hand, if the analysis is perpendicular to the polarization direction of the excitation, the expression for the intensity becomes

$$\mathscr{I}_{\perp} = (\alpha^2_{XY}A + \alpha^2_{XZ}B)(2C_0 + C_2) + (\alpha^2_{YY}A + \alpha^2_{YZ}B)C_2$$

$$+ (\alpha^2_{ZY}A + \alpha^2_{ZZ}B)4C_1 .$$

(64)

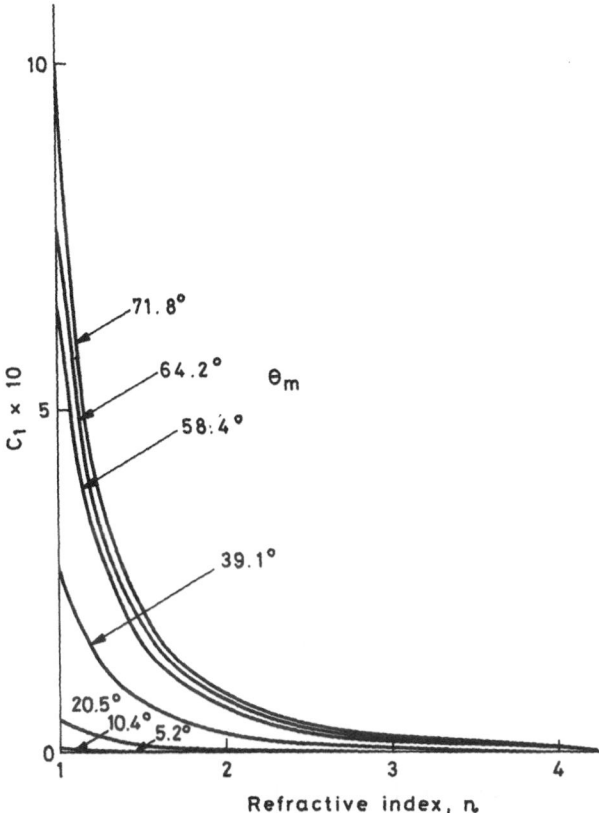

Fig. 2.17. Parameter C_1 as a function of refractive index [See Eq. (62).]

In terms of the tensor invariants defined by Eqs. (6–8), the scattered intensities are given by

$$\mathscr{I}_{\parallel} = \left[\frac{1}{3} A\Sigma^0 + \frac{1}{6} B\Sigma^1 + \left(\frac{2}{15} A + \frac{1}{10} B\right)\Sigma^2\right] 2C_0$$

$$+ \left[\frac{1}{3} B\Sigma^0 + \frac{1}{6} A\Sigma^1 + \left(\frac{1}{10} A + \frac{2}{15} B\right)\Sigma^2\right] 4C_1$$

$$+ \left[\frac{1}{3} A\Sigma^0 + \left(\frac{1}{6} A + \frac{1}{3} B\right)\Sigma^1 + \left(\frac{7}{30} A + \frac{1}{5} B\right)\Sigma^2\right] C_2 \qquad (65)$$

and

$$\mathscr{I}_{\perp} = \left(\frac{1}{6} \Sigma^1 + \frac{1}{10} \Sigma^2\right)(A + B)\, 2C_0$$

$$+ \left[\frac{1}{3} B\Sigma^0 + \frac{1}{6} A\Sigma^1 + \left(\frac{1}{10} A + \frac{2}{15} B\right)\Sigma^2\right] 4C_1$$

$$+ \left[\frac{1}{3} A\Sigma^0 + \left(\frac{1}{6} A + \frac{1}{3} B\right)\Sigma^1 + \left(\frac{7}{30} A + \frac{1}{5} B\right)\Sigma^2\right] C_2, \qquad (66)$$

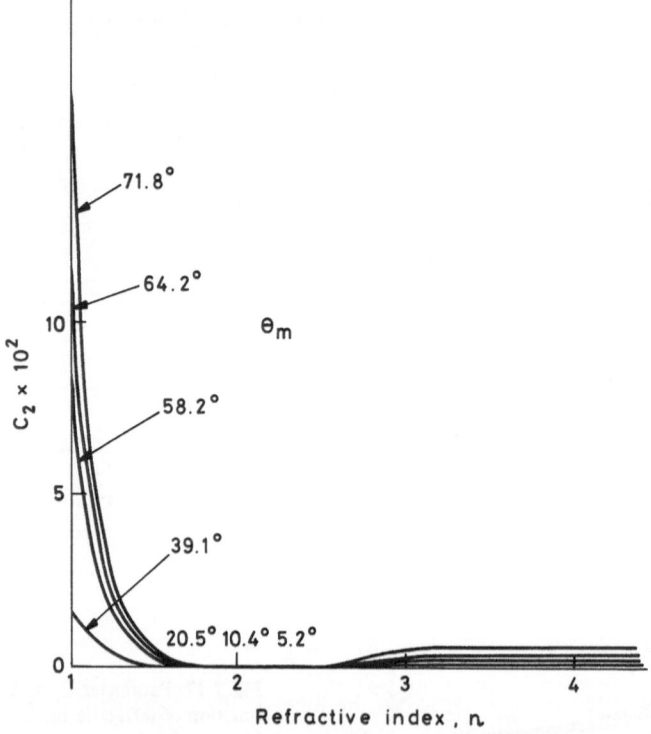

$C_2 \times 10^2$

71.8°

64.2°

10

θ_m

58.2°

5

39.1°

20.5° 10.4° 5.2°

0

1   2   3   4

Refractive index, n

Fig. 2. 18. Parameter C_2 as a function of refractive index [See Eq. (62).]

respectively. Equations (65) and (66) are then used to calculate the depolarization ratio defined by $\varrho = \mathscr{I}_\perp / \mathscr{I}_{||}$, as given by Eq. (16).

The analysis presented above has been employed in the interpretation of Raman spectra of both isotropic and anisotropic samples. In the case of isotropic samples correct depolarization ratios can be obtained, even with the use of high numerical-aperture objectives [22, 23]. However, if a beamsplitter is included in the optical system, its transmission characteristics must be evaluated and appropriate corrections introduced (see Chap. 6 of this book).

Optical anisotropy complicates the analysis considerably. However, the depolarization effects introduced by wide-angle objectives is still not large, and can be evaluated theoretically, if the birefringence of the sample does not play a significant role [24]. The depolarization due to the highly convergent incident and divergent scattered light is especially important for propagation in a direction close to that of an optical axis of a crystalline sample. This effect can be minimized by reducing the optical path (depth of focus) within the sample [24].

The application of the above analysis in resonance-Raman spectroscopy has also been demonstrated. In this case the nonvanishing of the invariant Σ^1 often results in so-called inverse depolarization, in which ϱ becomes very large [23].

2.9 Films and Adsorbed Species

Early in the laser era of Raman spectroscopy it was shown that it was possible to observe the spectra of molecules adsorbed on surfaces. It soon became evident that films of even monomolecular thickness could yield useful Raman spectra. The evolution of this area of research was due primarily to the introduction of advanced instrumentation. The most important contributions were initially the development of double and triple monochromators, highly sensitive detectors — including multi-element arrays — and efficient data-handling systems. The very recent introduction of Fourier-transform methods will most certainly result in further progress in this important application of Raman spectroscopy.

Numerous sampling systems for the study of thin films have been described. However, the basic geometries are essentially of two types: The backscattering configuration described earlier (Sect. 2.7) and a glancing-incidence arrangement, as represented in Fig. 2.19. The former system was employed in early investigations of the Raman spectra of various adsorbed species [26]. The commercialization of instruments of the type MOLE [27] greatly facilitated work in this general area, although the backscattering geometry, which is characteristic of these instruments, is more useful in the study of relatively thick films (>10 μm), particularly if normal incidence is employed.

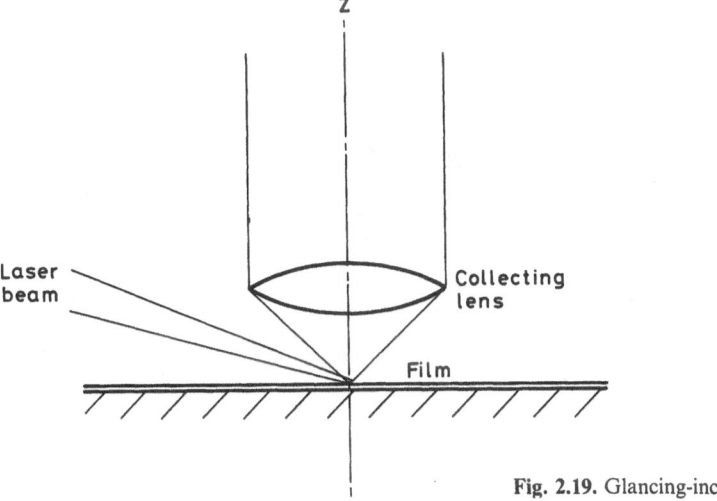

Fig. 2.19. Glancing-incidence geometry

For the Raman spectroscopic investigation of thinner films, a method which was initially proposed by Greenler and Slager [28] was applied to the analysis of thin films deposited on metal surfaces. In their experiments the Raman spectra of thin layers of benzoic acid deposited on a silver film were obtained with the sampling system shown in Fig. 2.20. In this case it was shown that the optimum angle of incidence for laser excitation is approximately equal to 70°, while the collection efficiency

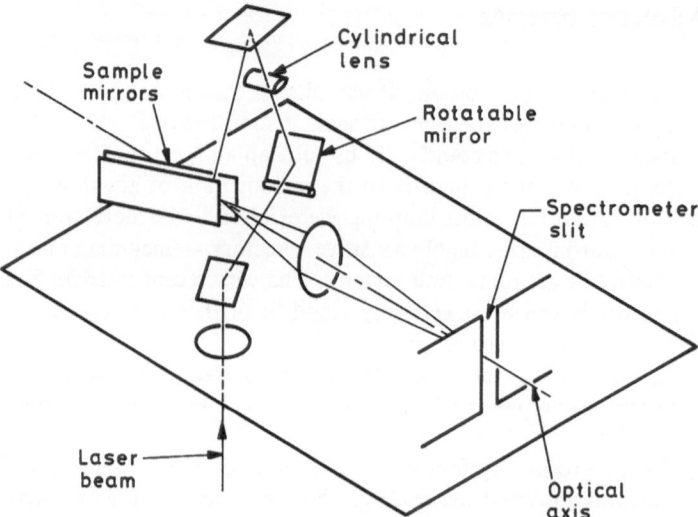

Fig. 2.20. Optical arrangement used in the investigation of films deposited on metallic surfaces [From Ref. 28]

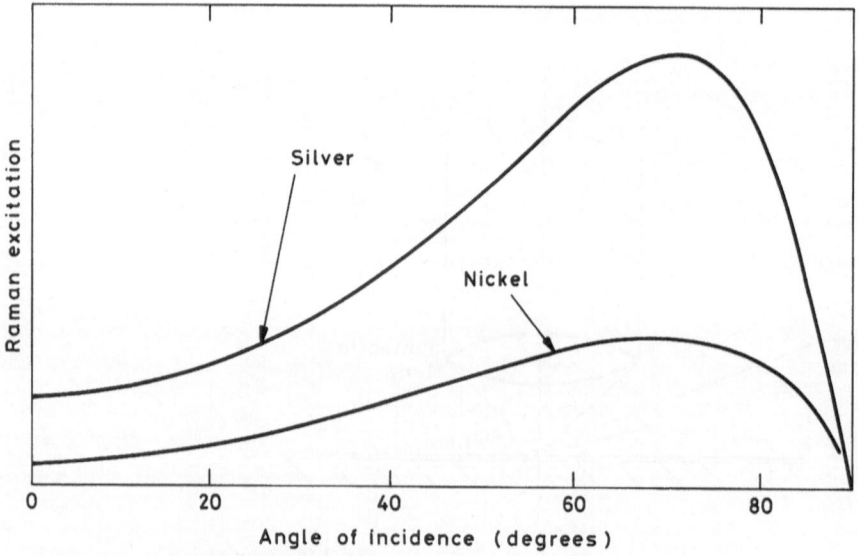

Fig. 2.21. Raman excitation as a function of angle of incidence [From Ref. 28]

is maximal at around 60° with respect to the surface normal. Under these conditions satisfactory Raman spectra of 50 Å deposits of benzoic acid were recorded. The optical geometry employed in these experiments was determined on the basis of the following analysis.

In an earlier paper by Greenler [29] it was demonstrated that the intensity of incident light near a reflecting surface is maximum at an angle far from the normal, whose

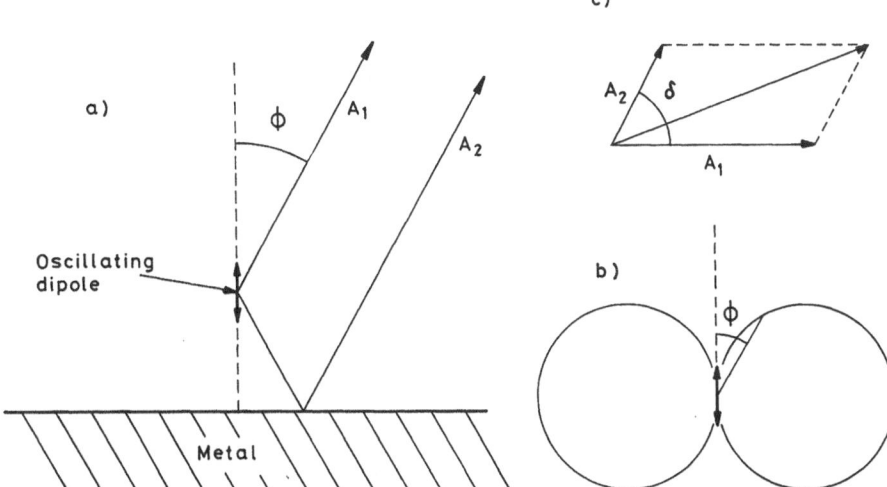

Fig. 2.22. Model of Raman-scattering intensity, a) Oscillating dipole near a metallic surface, b) Radiation from an oscillating dipole as a function of angle φ, c) Addition of amplitudes of direct and reflected rays [See Eq. (67).]

value depends on the optical constants of the surface. The calculated results for Ag and Ni surfaces are shown in Fig. 2.21 for light polarized parallel to the plane of incidence. It was further shown that perpendicularly polarized light results in a negligible intensity near the surface. It was also assumed in the analysis that the angular distribution of Raman scattering by a molecule near a metal surface can be represented by the radiation of an induced dipole oscillating perpendicular to the surface. If the distance between the dipole and the metal surface is small compared to the wavelength of the emitted light, the amplitude of the emission at an angle φ with respect to the surface normal can be calculated as shown in Fig. 2.22. Thus, if A_1 and A_2 are the respective amplitudes of the direct and reflected light ray, the intensity is given by

$$\mathscr{I} \propto A_1^2 + A_2^2 + 2A_1A_2 \cos \delta , \tag{67}$$

where δ is the reflection phase shift. If the amplitude of the light emitted perpendicular to the dipole is taken to be equal to unity, the relative amplitudes of the emitted and reflected rays become $A_1 = \sin \varphi$ and $A_2 = r \sin \varphi$, respectively, where the values of φ and r can be determined from the optical constants of the metal surface. The necessary relations, which were developed by Born and Wolf [30], are

$$\tan \delta = \frac{2b \cos \varphi(a^2 + b^2 - \sin^2 \varphi)}{a^2 + b^2 - (n^2 + k^2)^2 \cos^2 \varphi} \tag{68}$$

and

$$r^2 = \frac{(n^2 - k^2) \cos \varphi - a^2 + 2nk \cos \varphi - b^2}{(n^2 - k^2) \cos \varphi + a^2 + 2nk \cos \varphi + b^2} , \tag{69}$$

where a and b are defined by

$$a^2 = \frac{1}{2}\{[(n^2 - k^2 - \sin^2 \varphi)^2 + 4n^2k^2]^{1/2} + n^2 - k^2 - \sin^2 \varphi\} \qquad (70)$$

and

$$b^2 = \frac{1}{2}\{[(n^2 - k^2 - \sin^2 \varphi)^2 + 4n^2k^2]^{1/2} - n^2 + k^2 + \sin^2 \varphi\}, \qquad (71)$$

respectively. Here n and k are the real and imaginary parts of the complex refractive index of the metal, which is defined by $N = n - ik$. The results of numerical calculations based on the above analysis are presented in Fig. 2.23. The optimum collecting angle for Raman scattering is near 60° for these examples.

The glancing incidence geometry has been applied in numerous studies of the Raman spectra of adsorbed species and of thin films deposited on reflecting surfaces. However, in some cases the more recent, integrated optical techniques, which are discussed in the following section, offer important advantages.

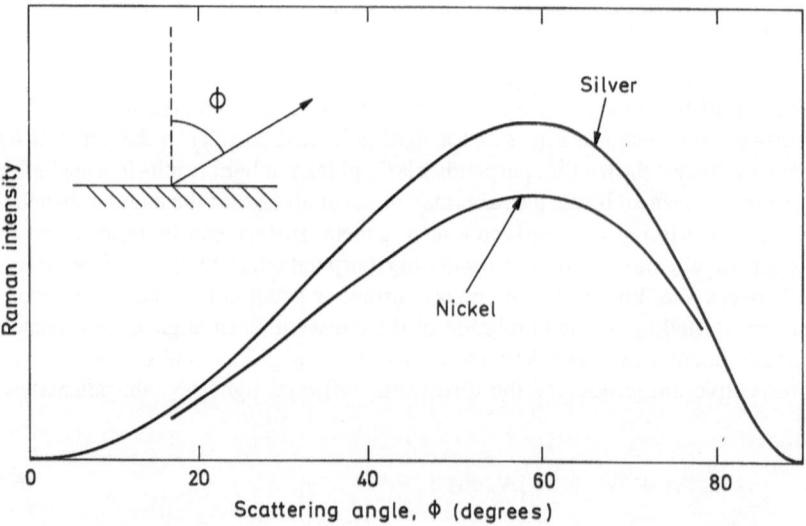

Fig. 2.23. Calculated relative Raman intensity as a function of scattering angle

2.10 Integrated Optics

A more sophisticated approach to the investigation of thin films by Raman spectroscopy has developed from the early work of R. Dupeyrat and coworkers [31]. The method involves an integrated optical system in which the sample film serves as a waveguide or a dielectric wall.

In the earliest experiments a thin-film sample of refractive index n_2 was deposited on a support, as shown in Fig. 2.24. The excitation was provided by a slightly conver-

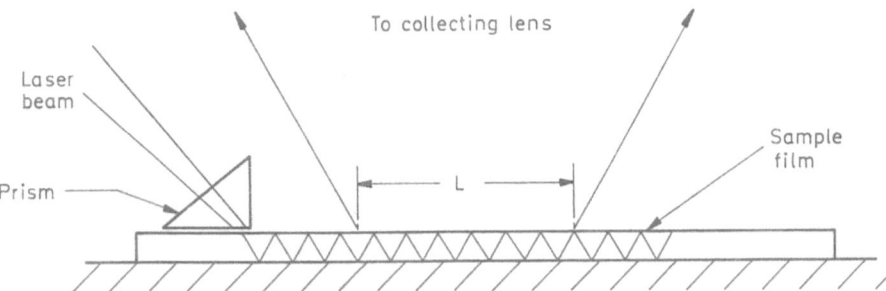

Fig. 2.24. Application of the waveguide technique to Raman studies of films

gent laser beam which entered the coupling prism through its hypotenus plane. If the angle of incidence and the polarization direction are properly chosen, the light will propagate as a guided wave in the sample film. The Raman-scattered light is usually collected perpendicular to the surface with the use of a microscope objective. The image of the entrance slit of the monochromator is adjusted so that its width corresponds to the diameter of the laser beam. Its length is equal to L, as shown.

The boundary conditions which determine the coupling of the excitation to the waveguide sample have been discussed in detail by Levy and Dupeyrat [32]. They showed that one or several modes of either the TE or TM polarizations can propagate, in a film of thickness σ according to the relation

$$2kn_2\sigma \cos \theta - \delta_{12} - \delta_{23} = 2\pi m , \qquad m = 0, 1, 2, ... \tag{72}$$

where $k = 2\pi/\lambda_0$, with λ_0 equal to the vacuum wavelength of the light; δ_{12} and δ_{23} are the phase shifts at interfaces 1–2 and 2–3, respectively, and θ_2 is the angle of reflection within the waveguide (Fig. 2.25). The phase shifts depend on the refractive

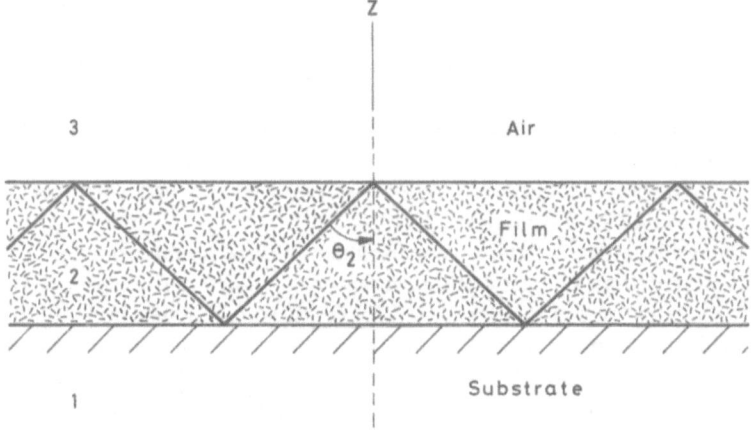

Fig. 2.25. Analysis of light propagation in a film deposited on a substrate

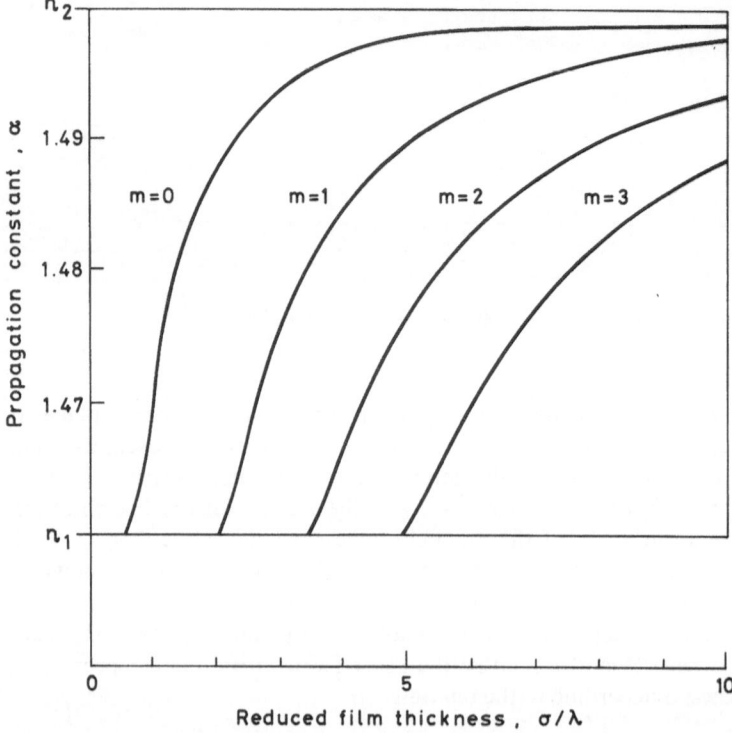

Fig. 2.26. Propagation constant, $\alpha = n_2 \sin \theta_2$, as a function of reduced sample thickness for TE modes. In this typical case $n_1 = 1.46$, $n_2 = 1.50$ and $n_3 = 1$

indices n_1 and n_3, and the propagation constant $\alpha = n_2 \sin \theta_2$ for a given polarization (TE or TM).

As an example, for the TE modes Eq. (72) yields the family of curves given in Fig. 2.26 for a typical case in which $n_1 = 1.46$, $n_2 = 1.50$ and $n_3 = 1$. With $m = 0$ it is seen that if σ decreases, α approaches n_1 and the energy penetrates more and more deeply into the support. On the other hand when σ increases, α tends towards n_2 and the energy is contained in the waveguide layer. In the latter limit the field of evanescent or exponentially damped waves at the exterior of the waveguide is zero. It is therefore possible by varying the sample thickness, to obtain separately or simultaneously the Raman spectrum of the waveguide sample or its support. The reported experimental results confirm this analysis.

Some additional advantages of the waveguide method were pointed out by Levy et al. [32], who estimated an intensity gain of a factor of ~ 2000 with respect to the backscattering geometry under similar sampling conditions. Furthermore, unlike the glancing-incidence geometry described in the previous section; the waveguide technique allows a choice of polarization directions. The TE modes can be excited by polarizing the laser beam in the Z, X plane and choosing the proper angle of incidence. Rotation of the polarization direction so that the electric vector is parallel to Y, accompanied by a small adjustment of the incident angle, will result in propagation of a TM mode. However, there is always a lower limit to the thickness of a film which can be used as a waveguide. In the above example it is equal to 0.56λ, as determined by the intersection of the $m = 0$ curve in Fig. 2.26 with the line $\alpha = n_1$.

In subsequent work by Dupeyrat et al. [33] a multilayer structure was employed which can be used to obtain the Raman spectra of thinner films. The analysis of this system is based on Fig. 2.27, in which the four layers are characterized by their respective refractive indèces and thicknesses. In this arrangement the exciting laser beam enters the system from below through medium 1, at an angle θ_1 with respect to the vertical (Z) axis. Media 2 and 3 are both thin films, while the semi-infinite medium 4 is usually air. The Raman scattering is observed as before, from above, with a microscope objective.

The parameters of the various media are chosen to produce inhomogeneous fields in media 2 and 4, and homogeneous ones in medium 3. The necessary conditions are realized for n_1, n_3 and n_4 real if θ_1 is larger than the critical angle defined by

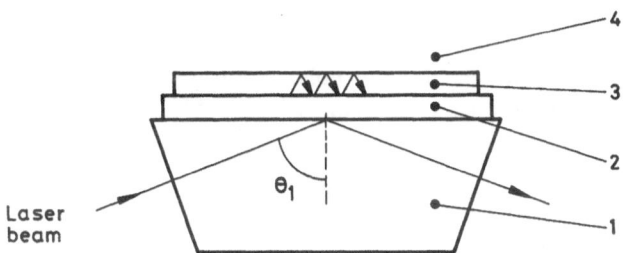

Fig. 2.27. The multilayer integrated optical system. Excitation is at an angle θ_1 with respect to the vertical axis

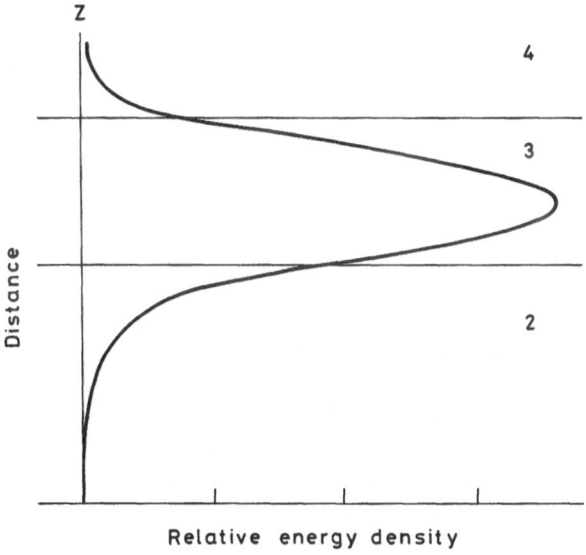

Fig. 2.28. Relative energy density as a function of vertical distance in a multilayer system

$\sin^{-1} (n_4/n_1)$. The resonant condition in the thin layer 3 is given by Eq. (72), where δ_{12} and δ_{23} are replaced by δ_{23} and δ_{34}, which represent the phase shifts at the boundaries 2–3 and 3–4, respectively. Note that the phase shift δ_{23} depends on the presence of medium 1, and, as σ_2 approaches infinity, the system becomes (aside from notation) equivalent to Fig. 2.25.

The energy density in layers 2, 3 and 4 relative to that in layer 1 is shown in Fig. 2.28 for a typical multilayer system. The parameters used are specified in the figure caption. It is apparent that a very large increase in energy density can be obtained in film 2. A calculation shows that the corresponding Raman scattering is approximately 3000 times greater than that obtained from backscattering measurements [32]. And, although the total Raman scattering is in this case about the same as in the waveguide method outlined above, the thickness of the film can be significantly reduced [34].

The integrated optical techniques outlined above have been developed and applied to a number of multilayer systems, in particular by Rabolt, Swalen et al. [34, 35]. Their group has described a sampling arrangement [36] which assures good optical contact between the coupling prism (as shown in Fig. 2.24) and the sample film over a wide temperature range. The experimental results obtained on various films of the Langmuir-Blodgett type are indicative of molecular orientation, and sensitive to order-disorder transitions. A detailed analysis of polarization measurements with the use of this sampling configuration has been given by Harrand [37].

2.11 Excitation with Circularly-Polarized Light

If two plane waves of the same frequency, which are propagating in the same direction, are combined, the polarization of the resulting wave depends on their relative amplitudes and phases. Thus, in the special case in which their amplitudes are equal and their phase difference is an odd multiple of $\pi/2$, the resulting wave is said to be circularly polarized. These two conditions can be expressed by [2]

$$\mathscr{E}_X^0 = \mathscr{E}_Y^0 \tag{73}$$

and

$$\frac{\mathscr{E}_X}{\mathscr{E}_Y} = \pm i, \tag{74}$$

respectively, for propagation in the Z direction. The sign in Eq. (74) determines the sense of rotation of the circularly polarized wave.

The terms "right-" and "left-handed" are usually employed in the literature to describe the sense of circular polarization. However, their meaning depends on the convention adopted. Physicists tend to follow Born and Wolf [30] for whom "right-handed polarization" specifies light whose electric vector and Poynting vector (in the direction of propagation) are described by a right-handed screw in time. On the other hand, chemists, who are accustomed to measurements of optical activity, use "right-handed polarization" if an observer looking in the direction from which the radiation is coming sees the end of the electric vector in rotation in the clockwise sense. However, it is evident that in the Raman application either convention leads to the same result, as long as optically-active samples are not being considered.

It is convenient in an initial description of Raman experiments with circularly polarized light, to assume a scattering angle of zero degrees, that is, that the exciting and scattered beams propagate in the same direction. It can be shown that the scattered intensities corresponding to corotation and contrarotation with respect to the excitation are given by

$$\mathscr{I}_{co} \propto \frac{1}{30} (10\Sigma^0 + 5\Sigma^1 + \Sigma^2) \tag{75}$$

and

$$\mathscr{I}_{contra} \propto \frac{1}{5} \Sigma^2 , \tag{76}$$

respectively. These expressions are obtained by averaging over the molecular orientations and are, therefore, applicable to the study of gaseous and liquid samples.

By analogy with the depolarization ratio, the reversal coefficient, R, can be defined by

$$R = \frac{\mathscr{I}_{contra}}{\mathscr{I}_{co}} = \frac{6\Sigma^2}{10\Sigma^0 + 5\Sigma^1 + \Sigma^2} \tag{77}$$

for the hypothetical zero-degree scattering angle. The reversal coefficient is then restricted to the range $0 \leq R \leq 6$. Furthermore, for ordinary Raman scattering, for which $\Sigma^1 = 0$, Eqs. (17) and (77) can be combined to yield

$$R = \frac{2\varrho}{1 - \varrho} . \tag{78}$$

In general, the reversal coefficient (unlike the depolarization ratio) is a function of the scattering angle and depends not only on R, as defined by Eq. (77), but also on ϱ. Thus, in principle, the measurement of the reversal coefficient with the use of circularly-polarized light at two different scattering angles is sufficient to determine both R and ϱ. However, certain special scattering angles are important. In particular, the conventional 90° configuration invariably yields a value of unity for the reversal coefficient. Furthermore, the backscattering experiment results in a reversal coefficient R(180°) = 1/R.

From an experimental point of view it is very inconvenient to attempt measurements at more than a single scattering angle. And, although the conventional 90° arrangement is perhaps the easiest for experiments with plane-polarized excitation, it does not, as indicated above, yield useful results with circularly-polarized light. In spite of other suggested possibilities (e.g., 60° [4]), most combined measurements of the depolarization ratio and the reversal coefficient have been made with the backscattering geometry.

The ratios Σ^0/Σ^2 and Σ^1/Σ^2 are related to ϱ and R by

$$\frac{\Sigma^0}{\Sigma^2} = \frac{(3/R) + 1 - 2\varrho}{5(1 + \varrho)} \tag{79}$$

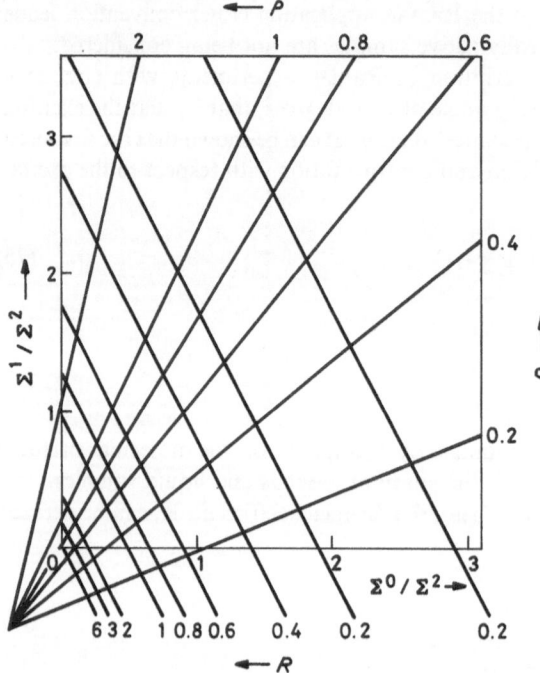

Fig. 2.29. Polarization nomogram for the 90° configuration (see text.)

and

$$\frac{\Sigma^1}{\Sigma^2} = \frac{6}{5R}\left(\frac{\varrho}{\varrho + 1}\right) + \frac{3}{5}\left(\frac{\varrho - 1}{\varrho + 1}\right), \tag{80}$$

which are obtained by combining Eqs. (17) and (77). Here ϱ is determined by direct measurement with the use of an analyzer and R is evaluated from the experimental value of $R(180°) = 1/R$. Equations (79) and (80) have been used by Mortensen and Hassing [4] to develop the nomogram shown in Fig. 2.29. One additional experimental observation must then be made in order to resolve the ratios Σ^1/Σ^2 and Σ^0/Σ^2, viz., the total scattered intensity obtained from Eqs. (10) and (11) in the form

$$\mathscr{I} = \mathscr{I}_{\parallel} + \mathscr{I}_{\perp} \propto \frac{1}{3}\Sigma^0 + \frac{1}{6}\Sigma^1 + \frac{7}{30}\Sigma^2 \tag{81}$$

$$= \frac{\Sigma^2}{30}\left[10\left(\frac{\Sigma^0}{\Sigma^2}\right) + 5\left(\frac{\Sigma^1}{\Sigma^2}\right) + 7\right]. \tag{82}$$

The last measurement, when combined with the values of the two ratios, as obtained from Fig. 2.29, yields a relative value of Σ^2.

2.12 Apologia

The author of any review article cannot hope to include all of the important developments in the area selected. Space limitations are paramount, while the choice of the material presented is determined primarily by the author's interests, and limited by his knowledge of the subject.

In the present chapter many important Raman spectroscopic sampling techniques have not even been mentioned, while others have not received their just due. In some cases excellent reviews have already appeared; it would be superfluous, then, to try to cover these subjects here. An extensive list of books and review articles on Raman spectroscopy is included at the end of this volume, in Chapter 7 Further Reading.

2.13 References

1. Placzek G (1934) Rayleigh-Streuung und Raman-Effekt. In: Marx E (ed) Handbuch der Radiologie. Akademische Verlag, Leipzig (vol VI,2), p 205
2. Long DA (1977) Raman spectroscopy. McGraw-Hill, New York
3. Wilson EBJr, Decius JC, Cross PC (1955) Molecular vibrations. McGraw-Hill, New York
4. Mortensen OS, Hassing S (1980) Polarization and interference phenomena in resonance Raman scattering. In: Clark RJH, Hester RE (eds) Advances in infrared and Raman spectroscopy, Heyden, London, vol 6
5. Edmonds AR (1960) Angular momentum in quantum mechanics. Princeton University Press, Princeton NJ
6. Kiefer W (1977) Recent techniques in Raman spectroscopy. In: Clark RJH, Hester RE (eds) Advances in infrared and Raman spectroscopy, Heyden, London, vol. 3
7. Gordon RG (1965) J. Chem. Phys. 43: 1307
8. Turrell G (1972) Infrared and Raman spectra of crystals. Academic Press, London
9. Damen TC, Porto SPS, Tell B (1960) Phys. Rev. 142: 570
10. Loudon R (1964) Adv. Phys. 13: 423
11. Loudon R (1965) Adv. Phys. 14: 621
12. Porto SPS (1969) Light scattering with laser sources. In: Wright GB (ed) Light-scattering spectra of solids. Springer, Berlin Heidelberg New York
13. Delhaye M, Migeon M (1966) C. R. Acad. Sci. Paris 262: 702
14. Delhaye M, Migeon M (1966) C. R. Acad. Sci. Paris 262: 1513
15. Barrett JJ, Adams NI III (1968) J. Opt. Soc. Am. 58: 311
16. Bridge NJ, Buckingham AD (1966) Proc. Roy. Soc. London A295: 334
17. Deb SK, Bansal ML, Roy AP (1984) Appl. Spectrosc. 38: 500
18. Turrell G (1985) J. Raman Spectrosc. 15: 103
19. Merlin JC, Delhaye M (1987) Raman and fluorescence investigation of biological samples with multichannel and micro-techniques. In: Stepanek J, Anzenbacher P, Sedlacek P (eds) Laser scattering spectroscopy of biological objects. Elsevier, Amsterdam, p 49
20. Strekas TC, Adams DH, Packer A, Spiro TG (1974) Appl. Spectrosc. 28: 324
21. Hendra PJ (1967) J. Chem. Soc. A: 1298
22. Brémard C, Dhamelincourt P, Laureyns J, Turrell G (1985) Appl. Spectrosc. 39: 1036
23. Brémard C, Laureyns J, Merlin JC, Turrell G (1986) J. Raman Spectrosc. 18: 305
24. Brémard C, Laureyns J, Turrell G (1987) Can. J. Spectrosc. 32: 70
25. Richards B, Wolf E (1959) Proc. Roy. Soc. London A253: 358
26. Hendra PJ, Turner IDM, Loader EJ, Stacey M (1974) J. Phys. Chem. 78: 300
27. Delhaye M, Dhamelincourt P (1975) J. Raman Spectrosc. 3: 33
28. Greenler RG, Slager TL (1973) Spectrochim. Acta 29A: 193

29. Greenler RG (1966) J. Chem. Phys. 44: 310
30. Born M, Wolf E (1965) Principles of Optics, 3rd ed. Pergamon, Oxford
31. Levy Y, Imbert C, Ciperiani J, Racine S, Dupeyrat R (1974) Optics Comm. 11: 66
32. Levy Y, Dupeyrat R (1977) J. phys. (col. C5) 38: 253
33. Cipriani J, Racine S, Dupeyrat R, Hasmonay H, Dupeyrat M, Levy Y, Imbert C (1974) Optics Comm. 11: 70
34. Rabolt JF, Santo R, Swalen JD (1984) Appl. Spectrosc. 34: 517
35. Rabe JP, Swalen JD, Rabolt JF (1987) J. Chem. Phys. 86: 1601
36. Barbaczy E, Dodge F, Rabolt JF (1987) Appl. Spectrosc. 41: 176
37. Harrand M (1986) J. Chem. Phys. 85: 2429, 2432

CHAPTER 3

Instrumentation for Raman Spectroscopy

by Don L. Gerrard and Heather J. Bowley
BP Research Centre, Chertsey Road, Sunbury-on-Thames,
Middlesex, England

In general terms the instrumentation required to undertake Raman spectroscopic studies is extremely simple. We require the following:
1. Some means of holding or containing the sample to be analysed.
2. A light source (laser).
3. A collection optic to collect the Raman scattered photons.
4. A monochromator to separate the Raman signal into its constituent wavelengths.
5. A detector to detect the photons at the various wavelengths where the Raman signal is produced by the sample and to give an output which is a measure of the relative intensities of the signals at these different wavelengths.
6. A computer system to make optimum use of the photons collected and to store and display the spectra.

These requirements will be considered separately, although when choosing a Raman system it is essential to consider the instrumentation as a whole and especially to consider it in the context of the nature of the work to be undertaken.

3.1 Sample Handling

The very high degree of versatility which the Raman technique offers in terms of sample handling is frequently overlooked by instrument manufacturers. After a slow start the technique of Raman microscopy for studying very small samples, is now well established and is catered for by manufacturers in the instrument design. This very important aspect of Raman spectroscopy is considered in detail in Chapter 6. On the whole, however, manufacturers do not yet pay enough attention to the other end of

Practical Raman Spectroscopy, Gardiner and Graves (Eds.)
© Springer-Verlag Berlin Heidelberg 1989

the problem — macro sampling. The technique of sampling is discussed in detail in Chapter 2, for gases, liquids and solids and so this section will just mention the use of some specialised hardware for sampling.

Because of the ability of visible lasers to penetrate even quite thick glass and the very weak Raman scatter of glass itself, it is possible to construct a wide range of specialist cells and associated hardware for the examination of systems not amenable to study by other analytical techniques. For example, the use of glass, often quite thick, as windows in high temperature/pressure cells means that reacting systems can be studied at a wide range of temperatures and pressures. In this way it is possible to undertake the Raman spectroscopic examination of chemical reactions under the same conditions as those used for large scale production. In order for this type of study to be successful it is necessary for the sampling area of the spectrometer to be sufficiently large to accommodate such cells, and it is in this respect that the manufacturers of commercially available instruments have been unimaginative, leaving the modifications of the spectrometer for this type of work to the individual user. There is no doubt, however, that one of the great advantages of the technique of Raman spectroscopy is its ability to be used for the spectral examination of systems under a

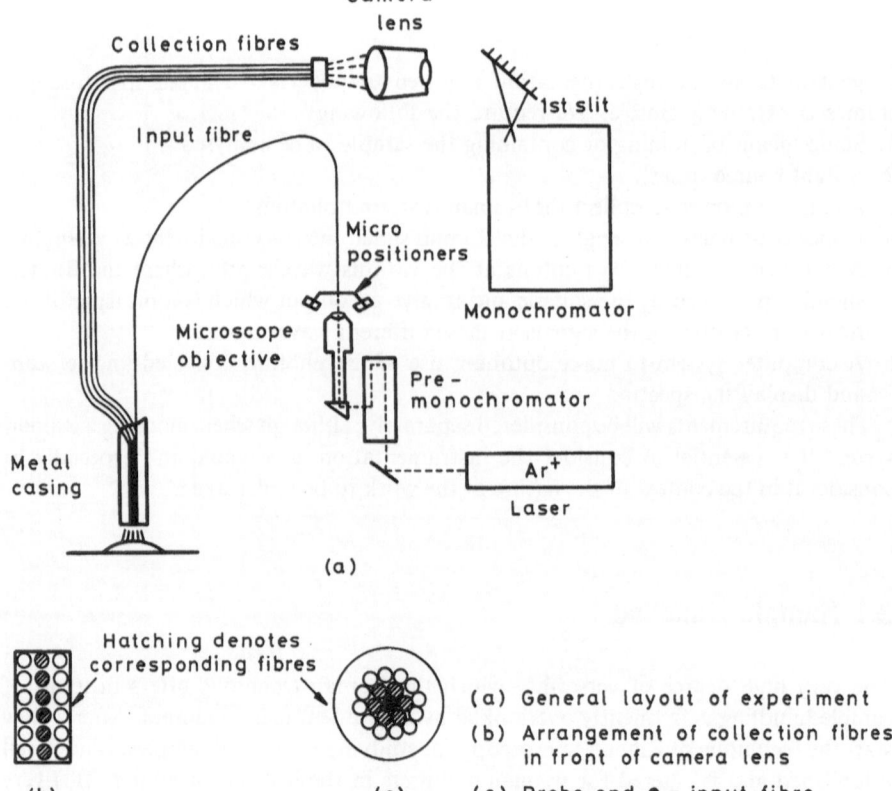

Fig. 3.1. Schematic diagram of a fibre optic probe for Raman spectroscopy

very wide range of temperatures and pressures with comparative ease, and in particular for the time resolved study of such systems. Specialised Raman cells are discussed in Chap. 5.

In the same context, another valuable feature of the high optical transmission of glass relates to the use of fibre optics [1]. It is a relatively simple procedure to focus the output from the laser beam into one end of a length of fibre optic (typically 200 μm in diameter) and the other end of the fibre can then be placed so that the emergent laser light illuminates the sample to be analysed. This then opens up the possibility of situating the sample remotely from the laser. Similarly since the Raman signal is normally in the visible region of the spectrum it is possible to collect the Raman scattered photons by means of one or more fibres placed close to the sample. The Raman signal can then be taken back via these fibres to the spectrometer and analysed in the normal way. Hence the sample to be analysed can be situated remotely from the spectrometer as well as from the laser. It is convenient in such a system to place the collection fibres round the single fibre which is carrying the laser beam, and in this way a simple fibre optic probe can be constructed. A schematic diagram of such a probe, showing the preferred orientation of the fibres is shown in Fig. 3.1 and is based on the design originally published by McCreery et al. [2].

The use of such fiber optic probes opens up the possibility of extending Raman studies to a wide range of systems which are not normally amenable to examination either by Raman or by other analytical techniques. For example, they can be used to study reactions occurring on, or close to, electrode surfaces. The probe end, sealed in a glass sheath can be placed directly into the electrolyte and the reaction monitored at an appropriate location. Also, the probe can be used to study samples in environments which are not suitable for use in the sample compartment of the spectrometer. This includes, for example, in-situ studies in molten salts or molten polymers, and in vivo biological studies. A spectrum of molten polyethylene obtained by placing the probe directly into the molten polymer is shown in Fig. 3.2. This contrasts with the spectrum of the polymer after solidification where there is a considerable degree of order, and this is reflected in the changes, particularly in the C-H deformation region, which are also illustrated in Fig. 3.2. Although this type of study could be undertaken in a special cell in the spectrometer sample compartment, such a system would only be able to show the changes occurring at the surface, whereas the fiber optic probe can show the effect deep in the bulk of the polymer, where cooling is much slower and where the final crystallinity will be very different from that at the surface.

Probes of the type described above can also be used in what can be called spectroscopically "hostile" environments. This includes, for example, situations where the material being examined may be highly corrosive, explosive, toxic or radioactive and special handling precautions have to be taken. There is also the long term possibility of using fiber optic probes to monitor large scale chemical reactions on the pilot plant or even production unit scale, where it could have implications for process monitoring. In this context the use of near infrared excitation in conjunction with a Fourier Transform Infrared spectrometer is especially interesting because this is a region of the spectrum where fiber optics are particularly useful and the technology well developed and the spectrometers used in this region lend themselves to being made sufficiently rugged for plant use.

Molten polyethylene

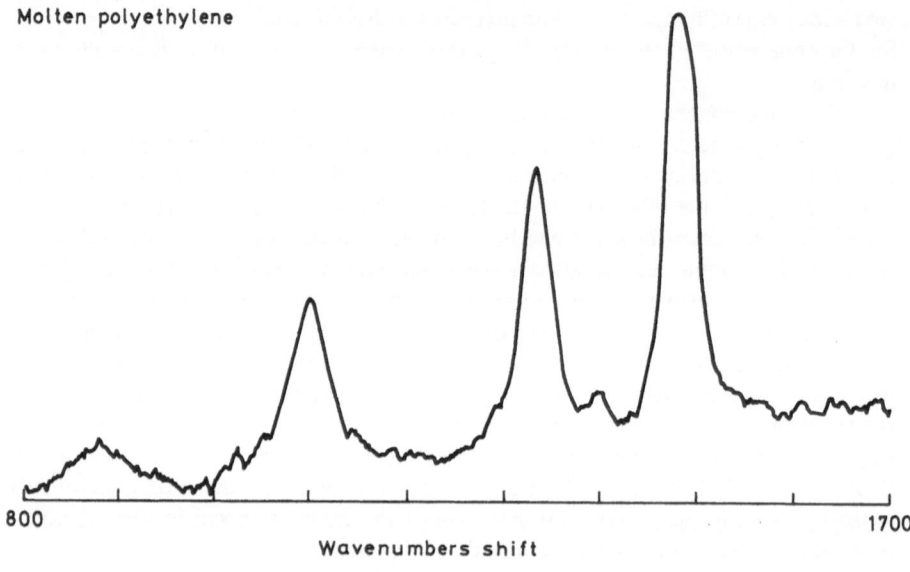

Wavenumbers shift

a

Polyethylene

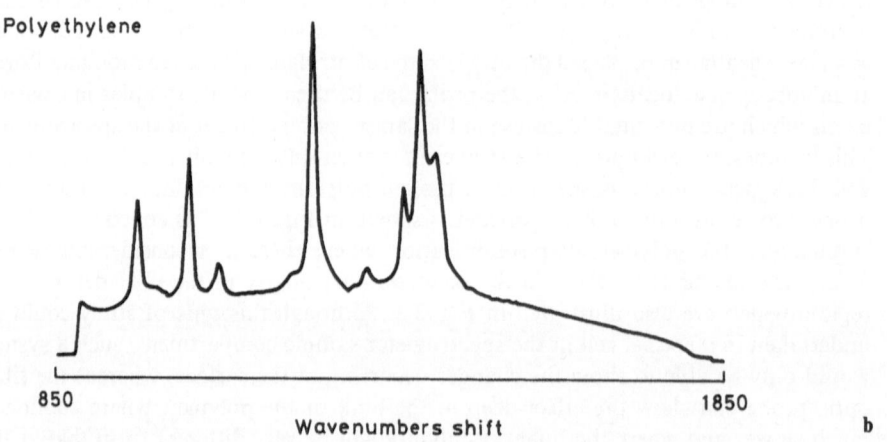

Wavenumbers shift

b

Fig. 3.2. Raman spectra of polyethylene obtained with a fibre optic probe: **a)** Molten polymer, **b)** Solid polymer

3.2 The Light Source

Although, with the rapid advances currently being made in light detection it may well be possible to obtain satisfactory Raman spectra from some materials using light sources other than lasers, there is no doubt that the laser, in its various forms, is currentl) the only excitation source that is realistically worth considering. The types of laser available fall conveniently into two categories, continuous wave (CW) and pulsed, and these will be considered separately.

3.2.1 Continuous Wave Lasers

As the name implies the continuous wave lasers give a continuous supply of photons and are by far the most widely used lasers for Raman spectroscopy at the present time. Initially, the most important of these, and the one which, to a large extent led to the revival of Raman spectroscopy as an analytical technique was the helium/neon laser, giving an output at 632.8 nm. Although this is still a useful laser, particularly because of its relatively low price, very robust nature, and the fact that it is air cooled, it has now been largely superseded by the much more powerful argon ion and krypton ion lasers. These give a series of lines, each of which can be used independently, and these are listed in Table 3.1. Some of these lines, particularly those in the near ultraviolet region of the spectrum are only available at a useful intensity from the most powerful version of these lasers.

On the whole the argon laser is of most value for studies requiring blue or green excitation and the krypton laser is of greatest value in the red and yellow regions of the spectrum. Of the two, provided that there is no specific wavelength requirement, the argon laser is probably more useful, and if funds are limited it is normally the better option. This is because it can be obtained in higher power output versions than the krypton laser, is less susceptible to instability due to pressure changes in the tube, and will give a wide range of lines without the need for changing the laser optics. These factors make the argon laser distinctly the easier of the two for the non-expert to use and in general it also tends to have less downtime than the krypton.

There are, however, other factors to be considered. For example, if the sample to be analysed exhibits significant fluorescence this may often be so severe that it totally obscures the Raman signal. This fluorescence can often be reduced considerably by using red excitation to obtain the Raman spectrum and for this purpose the 752.5 and 799.3 nm lines available from the krypton laser can be particularly useful [3]. Similarly, it may be the case that the samples to be examined are susceptible to photo-decomposition or thermal damage due to absorption at the wavelength of the laser line being used to obtain the Raman spectrum. This is much less likely to be the case with the lower energy red excitation. If possible, therefore, a broadly based Raman analytical laboratory should be equipped with both argon and krypton CW lasers. If the group undertaking the Raman work is fortunate enough to have more than one

Table 1. Wavelengths Available from Argon and Krypton Ion Lasers (nm)

Ar$^+$	Kr$^+$	Ar$^+$	Kr$^+$
528.7	799.3	454.5	476.2
514.5	793.1	379.5	468.0
501.7	752.5	363.8	415.4
496.5	676.5	351.4	413.1
488.0	647.1	351.1	406.7
476.5	568.2	335.9	356.4
472.7	530.9	334.5	350.7
465.8	520.8	333.6	337.5
457.9	482.5		323.9

Raman spectrometer, then a 25 mW helium/neon laser is also a valuable asset because of its ease of use, reliability and portability. The wide range of lines available from the argon and krypton lasers together is particularly useful if the work to be carried out involves resonance Raman studies, where it is useful to have available as wide a range of wavelengths as possible.

In general the power of the laser purchased should be as high as possible for several reasons. Firstly, the power inevitably reduces as the laser tube ages, and after three of four years in use it may well be down to only half of its original output power. Secondly, if a high power laser is available, then by careful planning of the positions of the laser and the spectrometer and the use of a beamsplitter it is possible to use the laser simultaneously on two spectrometers. Thirdly, a high power CW laser can be used in conjunction with a dye laser (see Sect. 3.2.4) to give a high degree of wavelength tunability where this is a specific requirement of the work to be undertaken. It must be remembered, however, that the CW laser tube only has a relatively short lifetime (typically about five years) even when great care is taken to ensure the use of very clean cooling water and to operate at power outputs well below the maximum. These tubes are expensive to replace and the higher the power rating of the laser the more expensive is the replacement tube.

Another type of CW laser which, with the advent of multichannel detector systems (see Sect. 3.5) is becoming increasingly attractive is the air-cooled argon laser. This can now be obtained in versions up to 100 mW (all lines) which for many purposes, is adequate for Raman spectroscopy [4]. These lasers, like the helium/neon, are robust, portable and relatively inexpensive.

A completely different type of CW laser, the CW YAG laser is a model which is of no use for conventional Raman spectroscopy, since it gives an output at 1.064 µm, well beyond the operating range of conventional monochromator/detector systems. It is now beginning to find application, however, in the area of Fourier Transform Raman spectroscopy, an exciting new field which is discussed in Sect. 3.4.2. The CW YAG laser has a high power output, is robust, aircooled, reliable and not expensive to maintain. In addition, by CW laser standards, it is relatively inexpensive. If the value of FT Raman spectroscopy is realised in the next few years then this type of laser will be a valuable addition to the Raman spectroscopist's hardware.

3.2.2 Pulsed Lasers

The use of pulsed lasers for Raman spectroscopy has, because of the very high powers involved, been largely in the area of nonlinear studies, and it is only relatively recently that their value in conventional Raman work has been appreciated. The two systems commonly used are based on either the pulsed YAG or excimer lasers. The YAG system relies on the fundamental output at 1.064 µm and its frequency doubled wavelength, 532 nm. The excimer uses a series of fundamentals, depending on the particular gas being used, in the ultraviolet region of the spectrum. Both of these lasers are normally used in conjunction with dye lasers and a range of frequency doubling crystals. The predecessor of the YAG or excimer system, based on the pulsed nitrogen laser is now scarcely used for this type of work because of the much greater versatility of the other two systems. The great advantage of these laser systems for the Raman spectroscopist is the very high degree of tunability which they offer. Versions are available for both systems which will give continuous tunability from about 190 nm to 4 μm. This is achieved by means of dye laser outputs and fundamentals being frequency doubled, tripled and mixed. A capability of this type opens up the possibility of extending resonance Raman and Surface Enhanced Raman Spectroscopy (SERS) to the study of a much wider range of compounds than is possible with CW lasers and also offers considerable potential for fluorescence suppression. The great advantage of both resonance and SERS effects from the analytical viewpoint is that they increase the sensitivity of the Raman technique and, in some cases its specificity. Since the number of organic compounds which exhibit electronic absorption spectra in the ultraviolet region is very much greater than in the visible region, the greatest advantage of these pulsed laser systems to the Raman spectroscopist is their tunability in the ultraviolet range.

Apart from the high degree of tunability, pulsed laser systems provide very high power outputs, often of the order of megawatts, but their main drawback lies in their very low pulse rate and the short duration of the pulses. The repetition rate for a YAG based system is typically 10 to 40 Hz and for an excimer about 50 to 300 Hz, and in each case the duration of the pulse is approximately 10 ns. This means that the detector is only exposed to the Raman signal for a maximum of 3 μs in every second, but in the rest of the time it is still building up a noise signal. To avoid the situation the detector needs to be gated so that it is only operating during the duration of the laser pulse. The most successful Raman system using tunable ultraviolet pulsed lasers is that described by Asher [5] which uses a gated intensified diode array detector and an ellipsoidal mirror as the collection optic and is based on a pulsed YAG laser.

Both the YAG and the excimer systems are extremely expensive to purchase and the excimer is expensive in day-to-day running costs. Both are intrinsically much more hazardous than the CW lasers because of their very high peak powers, and they also require appreciably more operator skill. In particular, their use to give tunable ultraviolet outputs needs very careful handling because of the obvious dangers associated with high power outputs of high energy photons where the beams cannot be observed by eye. Such systems are only of value, therefore, for highly specific applications where their tunability makes them preferable to CW lasers.

Another feature of the excimer and YAG pulsed lasers is that they can be made to give output pulses of less than 1 ps in duration. Again, it requires considerable expertise to keep such systems operating and their limited value to most Raman groups means that they would normally only be considered for specialist use by a very well equipped group. One of the most important potential advantages of such short pulses is their application in the area of fluorescence rejection [6]. As mentioned above, it is often the case that the fluorescence signal produced by the interaction of the laser beam and the sample being analysed is so severe that it totally obscures the Raman signal. This problem can frequently be overcome by using excitation in the far red [3] and near infrared [7], or in the ultraviolet [8]. Far red excitation can be obtained from a krypton CW laser, near infrared from a CW YAG laser, and ultraviolet from a pulsed YAG or excimer system. However, some materials still give fluorescent or thermal backgrounds even when the whole range of excitation wavelengths is used. In such cases the uses of sub picosecond pulses offers the hope of providing a solution. This relies on the fact that the rise time of the Raman signal is much shorter than that of the fluorescence. Hence, by using a very short pulse it is possible, at least in theory, to obtain the Raman signal produced by that pulse before a significant amount of fluorescence has been produced. Some success has been achieved using this technique with a gated diode array as the detector, although at the present time the method is limited by the gating systems available which are not capable of operating on a sub picosecond timescale. The sort of system to which this type of fluorescence rejection approach might apply are oils and oil-based products, commercial polymer samples and catalysts.

The pulsed lasers, at the present time, find their application in Raman spectroscopy largely in the area of resonance studies, especially of aromatic compounds where they have been used with considerable success in the study of hydrocarbons, phenols and biological systems. In due course they should also be used to extend the range of metal surfaces which can be used for SERS studies. It is unlikely that they will be standard items of equipment for the broadbased Raman group for some time because they are not generally applicable in the same way that CW lasers are, and are much more expensive and difficult to use.

3.2.3 Quasi CW Lasers

As mentioned above, one of the main drawbacks with the pulsed YAG and excimer systems is their low repetition rate which means that the detector, which ideally should be of the multichannel variety, is only exposed to the Raman signal for a very small proportion of the total length of time taken to accumulate the spectrum. Hence for most Raman studies CW lasers are used. Another alternative to be considered if high powers are required is the metal vapour laser, based on copper or gold which give pulsed outputs at a repetition rate of several kHz and are considered as quasi CW lasers. Between them, these copper and gold vapour lasers provide a range of lines in the visible region of the spectrum and give outputs equivalent to as much as 40 W from a CW laser. Air-cooled versions are available which are capable of delivering up to 10 W of output power. They can be used to pump dye lasers, but from the point of view of the Raman spectroscopist their main disadvantage is their large beam diameter (~ 1 cm) which is not very easy to reduce to a realistic diameter to make it suitable for most Raman purposes and in particular does not lend itself to use with a microscope attachment.

Another interesting quasi CW system is that described by Gustafson [9] based on two CW argon lasers and two dye lasers. This system is capable of producing tunable ultraviolet radiation down to 270 nm at a very high repetition rate and would provide an alternative to the excimer and YAG based systems in the ultraviolet region. Its drawbacks are that it does not operate at shorter wavelengths than 270 nm, which rules out the possibility of resonance studies of a wide range of organic materials, and it requires even more skill to set up and operate than the other two tunable UV lasers.

3.2.4 Dye Lasers

In principle the dye laser is extremely simple. Essentially it comprises a narrow jet of a dye solution through which a laser beam passes. This laser beam is the pump laser. This produces a broad band coherent output from the dye jet of longer wavelength than that of the pump laser. The efficiency of these systems is usually low. The broad band output can be tuned, using an optical wedge or birefringent filter and in this way the required monochromatic wavelength can be obtained. The pump laser can be one of the pulsed systems described above, or the violet output from a krypton CW laser or one of the higher power lines from an argon CW laser. In practice, for most purposes the high power CW argon laser with a boosted output in the ultra-violet is probably the best pump laser as it can be used to give tunability over the whole of the visible region of the spectrum, and even into the near infrared. To cover the whole of this region requires the use of several dyes, each of which only covers a relatively narrow range. For a comprehensive list of laser dyes and the wavelength ranges which they give, the reader is referred to the book by Maeda [10].

On the whole the power output that can be obtained over the visible region of the spectrum from a dye laser, pumped with a CW argon laser, is sufficient for Raman work, but is not sufficient for frequency doubling and mixing. The higher powers of the pulsed systems produce dye laser powers which are suitable for such purposes, and this accounts for their much greater wavelength range. Dye lasers in general are only of value if a significant amount of resonance work is to be undertaken when they can be used to match the wavelength of the appropriate electronic transition. Also in laboratories equipped only with an argon laser they provide a relatively inexpensive alternative to the krypton laser for use in the yellow and red region. However, they do require rather more operator skill and attention than the standard CW lasers.

3.3 Optics

Closely connected with the choice of laser is the choice of external optics associated with the spectrometer. These comprise, apart from any required to steer or adjust the height of the laser beam, the filtering system for CW lasers and the photon collection optics. It is an unfortunate feature of CW lasers that, associated with any of the individual laser lines is a series of other outputs, the plasma lines [11]. These are very much weaker than the laser output, but are comparable with, or slightly stronger than, typical Raman bands. Hence they can be confusing or can obscure the Raman signal in some cases. They appear as sharp bands in the recorded Raman spectrum and for the sake of convenience are normally removed before the laser beam comes into contact with the sample to be analysed. This can be done by spatially resolving the laser beam using a series of pinholes, but is more conveniently done using a filtering device. This can be one of two different types. The first is a conventional filter which will transmit the wavelength required at an acceptable intensity (typically about 50%) while removing the plasma lines, or at least reducing their intensity to a manageably low value. Depending on the laser or the wavelength being used a reduction in intensity of 50% or more may be unacceptably high. Only slightly more expensive, but a much more useful alternative is the pre-monochromator. This disperses the light coming in (i.e. required wavelength plus plasma lines) using a series of prisms. The dispersed light is then passed through an aperture sufficiently wide to take the light of the wavelength required but too narrow to take at the same time, the plasma lines which are significantly spatially separated from the wavelength of interest. The laser line is then focussed, by means of a lens, onto the sample.

Pre-monochromators of this type are tunable, normally using a vernier device and will operate over the whole of the visible region of the spectrum. Ultraviolet versions are also available, and these use quartz prisms. They have a high throughput (typically 75–80%) are extremely efficient at removing plasma lines and are robust and readily transferred from one laser to another. After removal of plasma lines it is usual to focus the laser onto the sample with either a cylindrical or spherical lens. The effects of using a spherical lens for this purpose are considered in Chap. 2, Sect. 2.4.

Another important feature of the optics external to the spectrometer's monochromator system is the collection optic. In the case of instruments which will be using visible radiation exclusively the normal collection optic is a camera lens. The massive input of effort in the area of demountable camera lenses in recent years and the intense competition has led to a wide range of very high resolution, high throughput lenses for the SLR camera market which are ideal as collection lenses for Raman spectrometers. Most commercial instruments come fitted with such a lens, but it is useful and by no means expensive to experiment with a range of such lenses of different focal lengths. This allows the sample to be at different distances from the collection optic whilst efficiently matching its focal length to the f number of the first mirror in the monochromator. Cassegrain mirrors are particularly good for samples positioned at a significant distance, but due to their central reflector, become inefficient at short working distances.

Because camera lenses are manufactured for purposes other than Raman spectroscopy their value is limited to the visible region of the spectrum. The Cassegrain mirror, however, can readily be obtained as a version suitable for working in the ultraviolet region. Asher has described a system which uses an ellipsoidal mirror to collect the Raman signal and this is an option which, for ultraviolet work, has proved extremely successful [5, 12–14].

If only one Raman spectrometer is to be used for a wide range of wavelengths including at least more ultraviolet work (if only from the CW lasers) then it is well worth having both Cassegrain and camera lens options available and to have quartz lenses throughout and to use optical surfaces coated for ultraviolet work. Although this will lead to some loss of throughput in the visible range, it rules out having to keep changing optics to work in the ultraviolet.

3.4 The Spectrometer

3.4.1 Conventional Dispersive Spectrometers and Spectrographs

The choice of monochromator is not quite as simple as it seems at first sight. Consideration must again be given to the sort of work that is to be undertaken. With the exception of a small amount of recently reported work which has used Fourier Transform Infrared spectrometers to record the Raman spectra, all Raman work is carried out using a dispersive monochromator of some kind. Before considering the relative advantages of the single, double, triple and single plus double monochromator system it is worth considering the differences between the spectrograph and the spectrometer. The grating (or gratings) of a spectrometer disperses the light entering the monochromator and then passes it through one or more narrow slits so that the light passing through to the detector at any one time has a very narrow band width and may be considered as essentially monochromatic. The spectrum is produced by rotating the gratings so that a continuously changing wavelength is presented to the detector. This is then displayed on a monitor, stored in the computer memory and plotted out directly onto an XY plotter. Scanning the whole spectral range of interest to the Raman spectroscopist (0 to 4000 cm^{-1}) takes, typically, 15 to 20 minutes. A spectrograph on the other hand uses much wider slits and a grating which produces much less dispersion and so gives a relatively broad band of light on a multi channel detector. These differences between the two devices give rise to advantages and disadvantages in each case.

The main advantage of the spectrograph is its high throughput and its main disadvantage is a relatively large spectral bandwidth reaching the detector at any one time, resulting in poor resolution and poor stray light rejection. On the whole, the Raman spectroscopist requires good resolution and stray light rejection and so it is most unusual to use a spectrograph in isolation for Raman studies. This means that spectrographs cannot normally operate within about a hundred cm^{-1} of the exciting line.

The main advantage of the spectrometer is its high resolution associated with a high degree of dispersion, and although the throughput is not as good as that of the spectrograph it is the type of instrument which has traditionally been used almost exclusively for Raman studies. Another important feature of the spectrometer, again resulting from its high degree of dispersion, is its good stray light rejection and this is an important consideration in deciding which spectrometer to choose — a single, double or triple. The single monochromator, using one grating is not sufficiently good in terms of stray light rejection or resolution for it to be worth serious consideration unless the work to be undertaken involves only extremely weak Raman scatterers, does not require spectra to be recorded in the vicinity of the exciting line and does not require particularly high resolution. It may well have applications with pulsed laser systems where the total photon count per second is likely to be extremely low and where its high throughput may be a more important feature than resolution or stray light.

The double monochromator using two gratings has for many years been the most widely used spectrometer for Raman studies. It combines a very high resolution with an adequate throughput for most purposes and has extremely good stray light rejection. This type of spectrometer, used in conjunction with a photomultiplier tube has, more than any other system, shown the value of Raman spectroscopy as an analytical technique and has established it as being worthy of consideration in its own right, not just as an expensive accessory to infrared laboratories. It has only been the advent of multichannel detectors (see Sect. 3.5) which has produced a new range of Raman instruments and which, for many purposes, are beginning to replace the conventional double monochromator system.

The true triple monochromator which briefly enjoyed some popularity, principally because of a particularly fine instrument produced by the now defunct Coderg company, offers slightly superior stray light rejection over the double system. This means that it can be used to record Raman bands that are very close to the exciting line (typically within a few wavenumbers). Its great disadvantage, however, is its very low throughput. This is bad enough in custom built triple spectrometers but in versions where the third element is available as an add-on to a conventional double system it has proved so bad that its application has been limited to the study of only very strong Raman scatterers. Spectra can often be obtained if a large number of individual spectra are accumulated but this is inevitably a time consuming process. The true triple spectrometer is now hardly ever used for Raman studies, its only realistic application being for samples when it is necessary to work very close to the exciting line, and there are relatively few materials of interest where this is necessary. If we are looking for a good quality, generally applicable spectrometer with a photomultiplier tube as the detector, which can work reasonably close to the existing line, has good resolution, and a throughput sufficient to obtain spectra of even very weak Raman scatterers than the double dispersive monochromator system is the ideal instrument. A schematic diagram of such an instrument is shown in Fig. 3.3.

The advent of multichannel detectors and their rapid development and performance improvement has once again focussed attention on the spectrograph type of instrument. This is because its ability to pass a relatively large wavelength range into the detector at any one time offers the capability of observing a significant portion of the total Raman spectrum of a sample at the same time without the need for scanning through the region of interest as is required with the conventional double monochromator/photomultiplier tube system. The spectrograph is therefore ideally suited to operation with a multichannel detector whereas the spectrometer, particularly

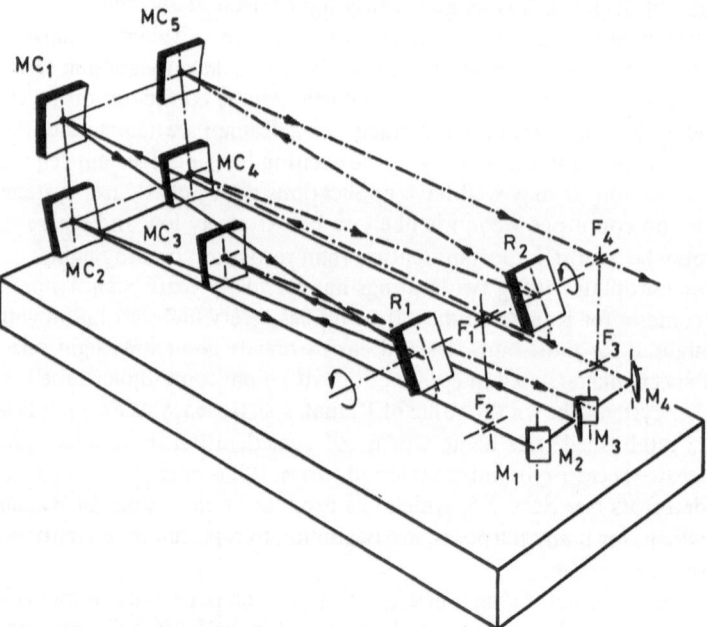

Fig. 3.3. Schematic diagram of a double monochromator. F_1 entrance slit, F_4 exit slit, MCn are concave mirrors, Mn are plane mirrors and Rn are plane gratings

a highly dispersive one, cannot take advantage of the ability of such detectors to observe a wide wavelength range. There arose a need, once multichannel detectors became accepted as a desirable addition to the Raman spectroscopist's hardware, for an instrument which would combine the advantages of the traditional double spectrometer (good resolution, high stray light rejection) with those of the spectrograph (good throughput, ability to observe large portions of spectrum) and this led to the development of a new generation of triple monochromators which are not triple dispersive instruments, but combine a double monochromator/spectrometer with a single spectrograph.

In most cases the double monochromator does not act in a traditional doubly dispersive manner, but as a double subtractive instrument. This means that the first element disperses the light in the normal manner, but the second element acts to recombine the light, after spatial filtering, before passing it through a slit into the spectrograph. This disperses the light and passes it on to the detector. In this way, the double element effectively acts as a very sophisticated filtering device for the spectrograph, by rejecting light at the excitation wavelength. The overall performance of this type of instrument is rather inferior to that of the traditional double dispersive spectrometer in terms of resolution, but this is more than compensated for by the ability to use a multichannel detector. On some instruments changing from one detector to the other is a very simple process as is the adjustment of the instrument parameters, such as slit width, associated with this change.

Triple grating spectrographs of this type have a great deal to offer, particularly to groups which are undertaking a wide range of Raman studies and do not specialise to any great extent in one particular area. They give the greatest possible versatility that can be obtained in any single instrument and are also, despite their many different functions, relatively easy to operate. They have grating mounts which enable a rapid change of gratings and thus enable coverage of a very large spectral range.

The current choice of Raman spectrometers for broad based analytical applications lies essentially between the double dispersive monochromator with a photomultiplier tube and the triple comprising a double subtractive monochromator with a spectrograph. The latter instrument could be fitted with both multichannel and photomultiplier tube detection. Of the two, the triple option is more versatile, particularly if dynamic systems are to be studied, but it is also considerably more expensive. The triple also lends itself better to pulsed work and can usually, if required, be used in its single spectrograph mode. If more accurate work on static systems is the main requirement than often the double would be the preferred choice.

3.4.2 Fourier Transform Raman Spectroscopy

Early attempts at Fourier Transform Raman spectroscopy were largely unsuccessful because of the weak nature of the Raman effect, and problems associated with removal of the exciting line wavelength from the scattered radiation made a considerable amount of the spectrum inaccessible. In this context, the conventional dispersive systems were far superior and no particular advantage was to be obtained by using Fourier Transform (FT) instrumentation, particularly after the widespread use of multichannel detectors which made dispersive instruments much faster. However, there is one area where the use of FT instruments is potentially very important and where several groups are now working and that is in the near infrared. Relatively inexpensive CW YAG lasers are now available which will give several watts at 1.064 μm. This region of the spectrum is potentially useful for the analysis of samples which exhibit fluorescence when exposed to the normal visible wavelength radiation from argon or krypton CW lasers or for samples which may exhibit photochemical

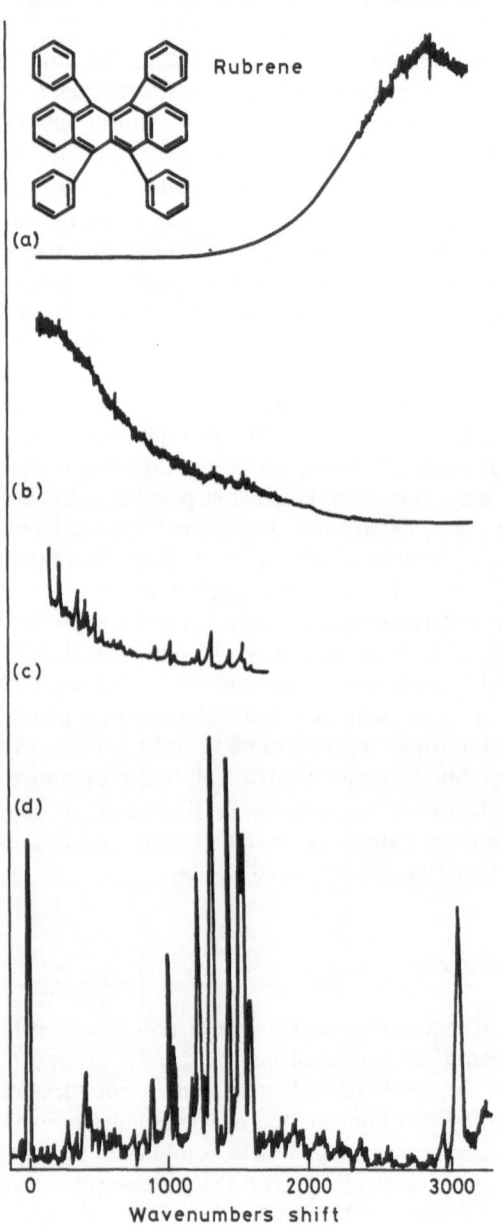

Fig. 3.4. Raman spectra of rubrene:
a) 514.5 nm excitation,
b) 632.8 nm excitation,
c) 752.5 nm excitation,
d) 1.064 μm excitation

or thermal decomposition due to absorption of these wavelengths. This fluorescence problem, as mentioned above, is the main reason why Raman spectroscopy is not yet widely used as a general spectroscopic analytical technique, especially in industry. Methods to reduce these fluorescence problems which have shown some promise have been mostly based on pulsed laser systems and the difficulties associated with the use of these lasers have been considered earlier (Sect. 3.2). If a simple method for overcoming fluorescence, based on a CW laser, could be devised it would prove a major step forward in the development and application of the Raman technique.

If the sample to be analysed is exposed to radiation at 1.064 µm instead of the more usual visible radiation, it will produce a Raman signal in the region 1.064 to about 1.85 µm, i.e. in the near infrared region of the spectrum. FT instruments are readily available for the detection of radiation in this wavelength range from several manu-facturers. In principle the Raman signal emanating from the irradiated sample is basically the same, from the FT infrared point of view, as an emission spectrum, which again can be routinely run on a conventional FTIR spectrometer. The only difference, and it is a very significant one, is the presence in the radiation of a very high intensity signal at 1.064 µm due to reflection and Rayleigh scatter. This radiation has to be removed before the detector in order to make the technique of using FTIR spectrometers to record Raman spectra a realistic possibility.

The basis of the method was first described by Hirschfeld and Chase [7] in 1986 and several groups have been active in the field since then. The hardware required to convert a conventional FTIR spectrometer for this purpose is relatively inexpensive and although it is unlikely that this type of system will replace dispersive mono-chromators for most applications they will undoubtedly be a very useful accessory to any vibrational spectroscopy laboratory. The value of this use of 1.064 µm excitation is illustrated in Fig. 3.4 which shows the spectrum of rubrene, a polycyclic aromatic hydrocarbon obtained with (a) 514.5 nm excitation from an argon ion laser, (b) 632.8 nm excitation from a helium/neon laser, and (c) 752.5 nm excitation from a krypton ion laser using a photomultiplier tube as detector and (d) 1.064 µm output of a CW YAG laser using an Indium Gallium Arsenide detector and a conventional FTIR spectrometer. In (a) the fluorescence is so severe that no bands are detectable and the further into the red the excitation wavelength goes the less this is a problem with the spectrum (d) being essentially fluorescence free.

The secret of this type of Raman instrumentation lies in having a detector which is as sensitive as possible in this region of the spectrum and in the removal of the 1.064 µm radiation as efficiently as possible. The latter stage can be achieved by using a stack of narrow band pass filters which will reduce the intensity of the exciting wavelength photons to an acceptable level without having a great effect on the Raman signal which in any case is much weaker in the near infrared than it would be for the same incident laser power in the visible. Even with a stack of four filters to remove the exciting line it is still difficult to obtain Raman signals within 100 cm^{-1} or so of the exciting line, but the value of the technique for facilitating Raman studies of fluorescent or photodegradable samples is potentially very great.

3.5 Detectors and Gratings

3.5.1 Gratings

Since the choice of gratings will be closely linked with the type of detector to be used, these two aspects of the instrumentation are considered together. The gratings will, to a large extent determine the resolution of the spectrometer. The more grooves per mm the better is the dispersion, which in the case of both photomultiplier tubes and multichannel detectors means greater resolution, although in the latter case it means that a more limited region of the spectrum will be monitored at any one time. If only one instrument is available and this is to scan over as wide a range as possible, but with reasonable resolution, the 1200 grooves per mm gratings are a good choice for a 1 m focal length monochromator. For better resolution but a rather smaller range the 1800 grooves/mm are extremely good and a combination of the two will enable operation over the whole of the visible region of the spectrum with good resolution. On modern instruments changing gratings is a relatively simple procedure. For very high resolution particularly in the ultraviolet region 2,400 or 3,600 grooves/mm are advisable and for a large spectral coverage with resolution supplied by the multi-channel detector, such as in a spectrograph, the 600 grooves/mm are a good option.

3.5.2 Photomultiplier Tubes

For many years the photomultiplier (PM) tube has been the only realistic option for Raman studies. Because of its high sensitivity, low background count, reliability and relatively low cost it is still the preferred detector for most applications. Specially selected tubes with very low background signals are readily available at a slightly higher price and virtually the whole of the visible and some of the ultraviolet region can be covered with one tube. The tubes can be used with any type of monochromator system but are of most value with double spectrometers when they are used to detect virtually monochromatic light at any one time. The spectrum is obtained by scanning across the range of interest and the continuous output from the tube fed into a chart recorder or displayed on a monitor.

The response of the tube is not linear over the whole of its range and this must be taken into account if spectra are taken over a wide wavelength range and especially if the exciting line lies towards one end of the operating range of the tube. This can be done by calibrating with a standard lamp or series of lamps (see Chap. 4 for how this may be done). The main disadvantage of the PM tube and a scanning double monochromator is the time taken to record the spectrum. For weak scatterers where a large number of accummulations needs to be used it may take several hours to record a reasonable quality spectrum. It also makes the study of dynamic systems difficult. For this reason the PM tube is best suited to static systems where no changes are occurring with time and it is of particular value for structural studies. If chemically or physically dynamic systems are to be studied the multichannel detector will normally need to be used.

3.5.3 The Intensified Diode Array

The intensified diode array, which to date is the most commonly used of the multi-channel detectors, has many apparent advantages over the PM tube and if these advantages can be related to the particular type of work to be undertaken then the diode array will be the preferred detector. A schematic diagram of an intensified diode array is shown in Fig. 3.5. Arrays comprise up to 1024 individual diodes, positioned in the exit focal plane such that each one is detecting at a slightly different wavelength, which enables a reasonably large portion of the spectrum to be displayed at any one time on a computer screen. A typical 1024 element array will cover a wavelength range of 26 nm. If the exciting line used to obtain the Raman spectrum is, for example, the 488.0 nm line of an argon laser this means that if the spectral scan is from a Δcm^{-1} of 100 upwards, this is in absolute terms 20,392 cm^{-1} or 490.4 nm. For an array covering 26 nm the upper limit of the spectrum being required is (490.4 + 26) nm, or 19,365 cm^{-1}. This is a total spectral range of 1027 cm^{-1} and so for a 1024 element array the limit of resolution imposed by the array, and ignoring any other factor is about 1 cm^{-1}. If on the other hand the 647.1 nm line of a krypton laser is used to obtain the spectrum then the starting point equivalent to a Δcm^{-1} of 100 is 15,354 cm^{-1} and a shift of 26 nm gives a finishing point for the spectrum of 14,765 cm^{-1} a total range of 589 cm^{-1}. This will improve the resolution, but will naturally reduce the wavelength range that can be displayed at any one time.

Another feature of the diode array detector is that, because of their extreme vulnerability to damage by high photon flux and the fact that they are normally used with spectrometers having a less than ideal stray light rejection, it is not normally possible to use them for studying bands close to the exciting line. Another disadvantage of the array is that it will deteriorate with time rather more rapidly than a PM tube, especially if it is regularly exposed to high photon counts and this deterioration will affect some diodes much more severely than others. In some cases this effect can be so marked that the relative intensities of different bands in the spectrum can be reversed merely by using different diodes to record them. This will make reliable quantitative measurements virtually impossible. In any case, great care has to be taken when carrying out quantitative work using a diode array detector (see Chap. 4 for a detailed consideration of this).

It is as well to ensure that if several spectra are to be obtained, for example in the study of a reacting system, that the same diodes are used each time to measure the same peaks. In this way differences between the responses of individual diodes can be eliminated. It is always necessary to undertake a regular check of the response of an array across all of its elements so that any deterioration can be taken into account. A simple way of checking the performance if the array is to take a strong scatterer with a sharp, single band, such as a diamond which gives a band at 1332 cm^{-1} and without altering the sample position or any of the instrument or laser parameters note the intensity of the peak at various positions on the array. Figure 3.6 shows the results for a defective multichannel plate. In this case the diamond band has been recorded at three different positions and the change in the recorded intensities across the array can be clearly seen. For an array in good condition and properly aligned there should be no more than a 1 to 2% differences in intensities recorded across the array. A response from a good array is also illustrated in Fig. 3.6. It is also necessary

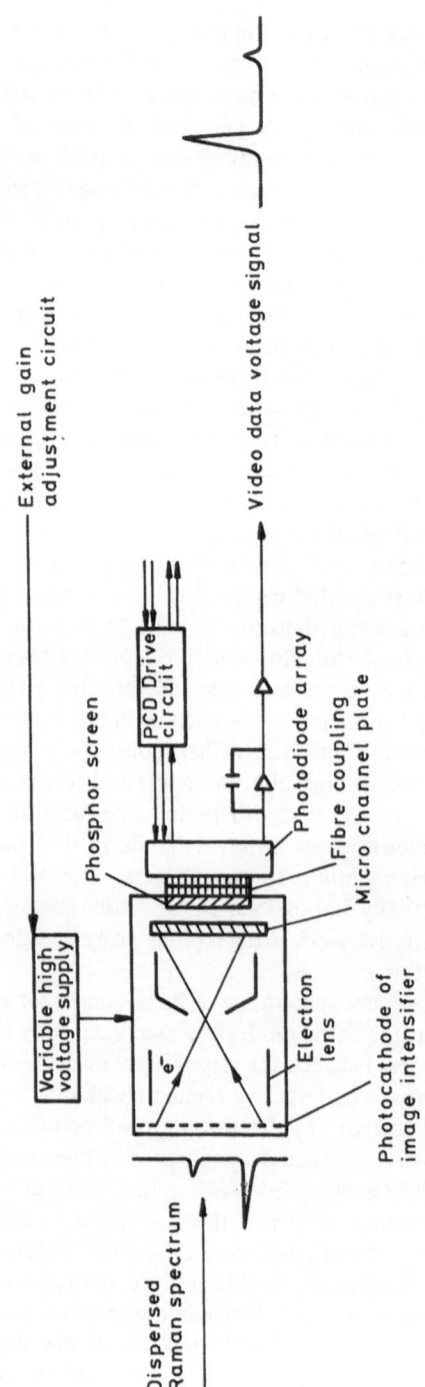

Fig. 3.5. Schematic diagram of an intensified diode array

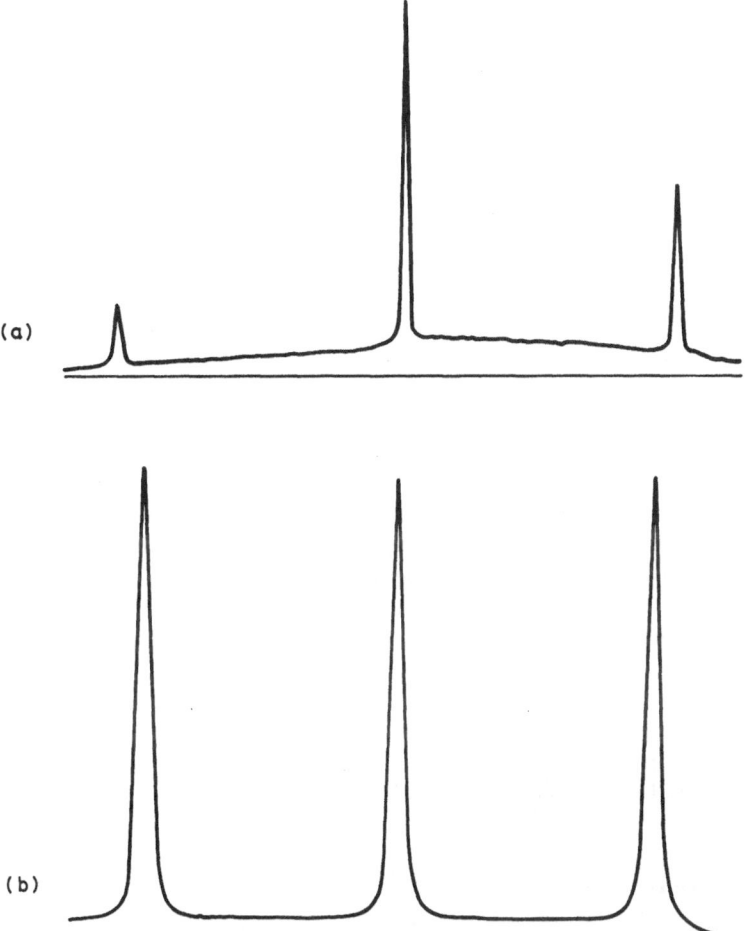

(a)

(b)

Fig. 3.6. Diode array response: **a)** Worn array, **b)** Array in good condition

to check the linearity of the response of the array — i.e. to ensure that by doubling the number of photons reaching the detector the recorded intensity also doubles. If it does not show good linearity this must be taken into account for quantitative work.

Despite the disadvantages of diode arrays there are enough advantages to make them worth serious consideration. Although the overall sensitivity at any one wavelength is, for most of the commonly used arrays, inferior to that of the PM tube the fact that a large portion of the spectrum is being monitored at any one time means that the time necessary to record a spectrum of any particular sample is normally much shorter than with the PM tube. Also, provided the background fluorescence is not too severe, sample alignment is very much easier with the array. The reason for this is that it is necessary with a PM tube to find a Raman peak, usually by rapidly scanning the spectrum and observing the response of the tube on a meter. The spectrometer is then slowly scanned over the peak and locked at the maximum while the various parameters such as sample position (x,y,z co-ordinates), slit width and laser power are optimised. If the sample is of unknown chemical composition it is quite easy, with a weak scatterer to miss the Raman peaks in a quick visual scan and the optimisation process can then be quite lengthy. With a diode array, however, the whole of the portion of the spectrum to be recorded is displayed on the screen, and in modern instruments this can be done every few milliseconds. So as far as sample alignment is concerned, the spectrum is displayed on what is virtually a real time basis. In this way, changes in the spectrum during alignment can be easily observed by eye.

To a large extent then the choice of a PM tube or a diode array depends, apart from financial constraints, on the nature of the work to be undertaken. If spectra need to be obtained rapidly for example, if the technique is to be used to study reaction mechanisms and kinetics, then the diode array is by far the more useful as it can give a good quality spectrum of the region of interest for many compounds in only a few seconds. The most accurate quantitative work is always best carried out with a PM tube which, once it is calibrated for its response over the spectral range in which it is to be used, will last for a long time without any significant deterioration. However, the extra work required in calibrating a diode array for quantitative work is well worth the effort because of the very much reduced spectral acquisition time.

If the Raman capability for the laboratory is to comprise only one spectrometer and a reasonably wide range of applications is to be addressed then a triple mono-chromator system (double plus spectrograph) fitted with both PM tube and diode array appears to be the best option. Such an instrument gives the capability of under-taking virtually any application of conventional Raman spectroscopy and is an extremely powerful and versatile analytical tool.

3.5.4 Charge Coupled Devices

Until very recently the only realistic alternatives for Raman detection systems were the PM tube and the diode array. A third possibility which appears to have consider-able potential and which is now being examined by several groups is the charge-coupled device (CCD) [15]. These devices have the ability to register almost every photon striking them, over a wide range of wavelengths and are now widely used in astronomy. For a detailed description of the mode of operation of CCD's the reader is referred to the paper by Janesrick and Bleuke [16]. The basic merits of the cooled slow scan CCD are the high resolution, high quantum efficiency, broad spectral response, large dynamic range, photometric accuracy and linearity.

They comprise a two dimensional array of PIXELS, typically up to 800 × 800 although systems up to 2048 × 2048 have been constructed. They can be used in two ways:

(a) For conventional Raman spectroscopy a proportion of the total numbers of PIXELS can be used, determined by focussing the image from the final slit onto the CCD.

(b) Of even greater potential interest is the use of CCD's for Raman imaging. In this application the Raman scatter from a portion of the sample is displayed on a screen. The spectrometer is used to pass only the particular wavelength of light we are interested in (i.e. the Raman peak for a particular species) onto the CCD and in this way the distribution of the species of interest can be imaged. This type of analysis has always been potentially one of the most valuable applications of Raman spectroscopy, but because of the lack of suitably sensitive detectors has only been applied to a very limited number of systems. It has obvious application in the area of catalyst studies, semiconductors, biology, mineralogy and polymers. Although the use of CCD detectors for Raman work is only in its infancy it may well provide the most exciting area of development in analytical Raman spectroscopy over the next few years.

3.6 Computing

The use of computers is obviously a very large subject and in this chapter we will be limited to discussing the minimum basic requirement for the computer system and also the various desirable capabilities that the Raman spectroscopist finds useful. For many purposes it is possible to obtain perfectly adequate Raman spectra with only a minimal computing capability. The laser parameters are normally controlled by hand, which leaves the monochromator and detector parameters to be considered. It is normal on modern instruments to use a computer to control the most crucial monochromator parameter, the slit width, opening and closing of the slits being achieved by means of stepper motors. However, this is only a matter of convenience, and all of the instrumental parameters can be easily adjusted by hand. Indeed, perfectly satisfactory Raman spectra of a wide range of materials can be obtained without any computing capabilities except, in the case of a diode array that required to operate the detector. Modern spectrometers are normally supplied with a dedicated computer and a software package to operate the spectrometer and to perform a range of operations on the acquired data. In this latter context the most important feature is the ability to accumulate spectra. The Raman spectra of many materials, are extremely weak, and it is only by the accumulation of a large number of spectra that useful Raman data can be obtained. Also, as mentioned before, many materials exhibit fluorescence when exposed to the laser beam, and even when this is not so severe as to totally obscure the Raman signal, it will give a strong background signal. For reliable quantitative work to be carried out, this background needs to be removed, and the spectrometer manufacturers approach this problem in a variety of ways.

The simplest approach is to use a point-by-point background simulation method. This involves moving a cursor over the background and using the computer to generate, from the cursor points, a simulated fluorescence background. More sophisticated methods use a combination of a range of mathematical functions to simulate the fluorescence signal. However, the nature of these fluorescence backgrounds is so variable that it is virtually impossible to match them even by a complex combination of such functions. In any case background subtraction does not solve the associated signal/noise problem.

Another desirable feature of Raman software is the ability to undertake spectral subtraction. As with infrared spectroscopy it is often the case that multicomponent systems are encountered and one or more components may need to be subtracted from the total spectrum in order to extract the spectrum of the required component. Chapter 4 describes how this may be done. Such spectral subtraction routines are widely available from either the manufacturers of the Raman spectrometer or from manufacturers of FTIR spectrometers. Indeed, it is quite feasible, if Raman and FTIR instruments are available in close proximity to each other to transfer Raman data to the computer of the infrared instrument and access the FTIR software which at the present time is much more highly developed than the corresponding Raman software.

Although not essential for most purposes the use of deconvolution techniques is becoming more widely used in Raman spectroscopy. It is of particular value in the study of reacting systems where, for example, the Raman bands being monitored can overlap to such a great extent that, without deconvolution, they cannot be separately quantified. One of the most successful and widely used of these techniques is that of Fourier Self-Deconvolution which is also widely used in infrared studies. This does not require particularly sophisticated or expensive computing capability and can either be incorporated into the software of the Raman instrument or can be used via a FTIR spectrometer where commercial software packages are readily available.

Although for complex time-resolved studies, particularly those involving chemically dynamic systems, relatively sophisticated computing is a distinct advantage, the technique of Raman spectroscopy can nonetheless be used very successfully for a wide range of applications with very little computing capability. In this sense, the Raman hardware still has many similarities with dispersive infrared instrumentation and the two techniques resemble each other very closely.

3.7 References

1. McCreery RL, Fleischmann M, Hendra P (1983) Anal Chem 55: 46
2. Schwab SD, McCreery RL (1984) Anal Chem 56: 2199
3. Williams KPJ, Gerrard DL (1985) Optics and Laser Technology 245
4. Bowley HJ, Gerrard DL (1986) Optics and Laser Technology 93
5. Asher SA, Johnson CR, Murtaugh J (1983) Rev. Sci. Instrum. 54: 1657
6. Howard J, Everall NJ, Jackson RW, Hutchinson K (1986) J. Phys. E. Sci. Instrum. 19: 934
7. Hirschfeld T, Chase B (1986) Appl. Spectrosc. 40: 133
8. Jones CM, Naim TA, Ludwig M, Murtaugh J, Flaugh PL, Dudik JM, Johnson CR, Asher SA (1985) Trends Anal Chem, (Pers. Ed) 4: 75
9. Gustafson TL, Roberts DM (1982) Optics Commun. 43: 14
10. Maeda M (1984) Laser dyes, Academic Press, Tokyo
11. Loader J (1970) Basic Laser Raman Spectroscopy, Heydon
12. Asher SA, Johnson CR (1984) Science 225: 311
13. Johnson CR, Asher SA (1984) Anal Chem 56: 2258
14. Asher SA (1984) Anal. Chem. 56: 720
15. Murray CA, Dierker SB (1986) J. Opt. Soc. Am. A 3: 2151
16. Janesick J, Bleuke M (1987) Sky and Telescope 239

CHAPTER 4

Calibration and Data Processing

by P. R. Graves
Materials Development Division, Building 393, Harwell Laboratory, Didcot
Oxfordshire. 0X11 ORA/UK.

4.1 Introduction

Raman spectroscopy has been revolutionised in the past 10 years by the advent of new photon detectors and the increasing use of microprocessors and minicomputers to perform Raman data collection and processing. The aim of this Chapter is to set out in a systematic fashion the important aspects of data handling. Digital computers are used exclusively today, therefore, the first consideration should be to establish how accurately analogue-to-digital convertors can represent real photon data and how this influences subsequent treatment of that data. The question of wavelength or frequency calibration is also relevant, particularly to Raman spectral data sets recorded using array detectors. The Raman spectroscopist has many favourite 'standard' materials that are used to check instrument resolution and accuracy, but measurements of instrumental throughput and the polarising effects of gratings and slits are often ignored. Fortunately, digital methods can facilitate the normalisation of individual spectra and it should now be possible to compare the spectra from different Raman instruments easily and successfully. Once a Raman spectrometer or spectrograph's complete instrument function has been measured, it becomes possible to apply very powerful data fitting, estimation and filtering techniques to extract as much information from a Raman spectrum as possible.

Specific problems that better data processing is helping to solve are: (i) dealing with poor signal-to-noise ratio bands, (ii) removing sloping backgrounds, which may be due to laser induced sample fluorescence, and (iii) removing Raman line broadening due to instrument function. There have been many digital techniques reported in the literature in recent years, some of which are significantly better than others. Very often, one algorithm is to be preferred to another, depending on how well the statistics of the Raman datasets approximate to known models for Poisson

Practical Raman Spectroscopy, Gardiner and Graves (Eds.)
© Springer-Verlag Berlin Heidelberg 1989

or Gaussian noise corruption, or how accurately the recorded Raman lineshapes match Lorentzian or Voigt profiles. The aim of this Chapter is to help the Raman spectroscopist to make an informed choice as to which method is most appropriate to his or her problem.

4.2 Analogue-to-Digital Conversion

Two basic types of photon detection equipment are used in Raman spectroscopy, either single channel or multichannel. The first, usually a photomultiplier tube, may be used in DC mode, producing a current signal which is proportional to the number of photons hitting the photocathode at any particular time, or in AC mode, yielding a current pulse at regular time intervals which can be described as a photon count rate. These current pulses can be analysed by means of their pulse heights to reject photocathode dark current and then counted individually. The second type can be broadly described as multi-element one- or two-dimensional detectors such as intensified photodiode arrays, charge coupled devices (CCDs) and silicon targets (SITs and ISITs). Digitisation of analogue signals from these two types of detector places different requirements on the performance of analogue to digital converters (ADCs). In the case of the first type, a signal conversion is required perhaps every 0.1 seconds, however, if we are to take full advantage of the scan speed of an array detector, a conversion would be required every few microseconds in order to read a 1024 element detector every 8 milliseconds!

In digital conversion, resolution is always traded against speed. Therefore a single channel detector signal can be easily converted with 16-bit accuracy, giving between 4 and 5 significant figures. However, a photodiode array signal can, at present, be converted with only 14-bit accuracy and some systems provide only 12-bit accuracy. The 14-bit accuracy gives 4 significant figures and 12-bit gives only 3. Signal averaging or extended data acquisition times will indeed give larger photon count values, but no extra information will be gathered outside of the number of significant figures supplied by the ADC.

Other features of high performance ADCs do not significantly distort Raman spectra. Linearity is seldom a problem. Commercially available converters will quote linearity to at least $\pm 1/2$ of the least significant bit over the full dynamic range. The most important source of electronic noise is probably the pre-amplifier, being comparable to the shot noise of the detector and the Raman signal in most cases. It can be seen from this very brief consideration of ADCs and detection systems that the type of ADC chosen should be matched to the detection system, which should in turn be governed by the type of Raman experiment to be carried out.

4.3 Frequency Calibration

The optical monochromators used in most Raman spectrometers fall into two categories. They are either designed to disperse light linearly in frequency (wavenumber) or in wavelength. In conventional single channel detection mode the final

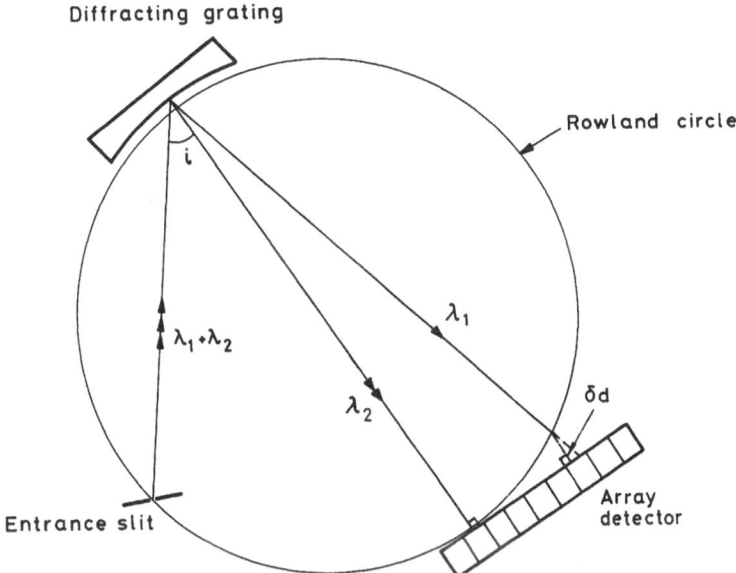

Diffracting grating

Rowland circle

λ_1

$\lambda_1 + \lambda_2$

λ_2

δd

Array detector

Entrance slit

Fig. 4.1. An array detector is positioned tangentially on the Rowland circle of the concave grating, so the grating equation holds for only one ray. The second ray does not impinge perpendicularly on the array therefore an apparent error in dispersion *delta* d is observed

grating or gratings are rotated by a cosecant drive or a sine drive. Sources of error in the frequency or wavelengths of such drives are therefore both absolute and relative. Absolute errors are due to inaccurate zero-ing of the spectrometer drive and random temperature fluctuations of the equipment. Relative errors in frequency or wavelength may be due to non-linearity in the drive bars (i.e. a variation in the pitch of the screw thread) or, more importantly in a multichannel detection system, how well the spectrometer design corrects the effect of placing a planar detector on the Rowland circle of a concave grating mount. Figure 4.1 illustrates this problem. The other major source of distortion of the focal plane that can occur in multichannel detectors is observed when an electrostatically focussed image intensifier is used in front of a photodiode array or CCD. Pin cushion distortion produces a 3% distortion in the focal plane, which should be corrected when standard lamp frequencies are fitted to observed peak positions. In this Section we consider what methods are available to the spectroscopist to calibrate Raman spectra with respect to absolute and relative systematic errors and give a measure of how accurate these calibrations are with respect to random errors.

The most obvious source of calibration frequencies for Raman spectroscopy is the plasma line emission spectrum from a gas laser. Normally these lines are removed with the use of a laser premonochromator to avoid obscuring weak Raman transitions of a similar frequency. Loader [1] first published a list of useful plasma line emission frequencies for argon ion lasers. However, the quoted values are not really sufficiently accurate and some line assignments must now be viewed as incorrect. Table 4.1 is a reprint of the results of Craig and Levin [2]. It should be noted that the relative

Table 4.1. Principal argon-ion laser plasma lines used for frequency calibration (taken from Craig and Levin [2])

Origin	Relative Intensity	Wavelength (air) (nm)	Wavenumber (vac) (cm⁻¹)	Shift (cm⁻¹) relative to 488.0 nm	Shift (cm⁻¹) relative to 514.5 nm
Ar$^+$	5000	487.9860	20486.67	0	
Ar$^+$	200	488.9033	20448.23	38.4	
Ar$^+$	130	490.4753	20382.70	104.0	
Ar$^+$	970	493.3206	20265.13	221.5	
Ar$^+$	14	494.2915	20225.33	261.3	
Ar$^+$	10	495.5111	20175.53	311.1	
Ar$^+$	960	496.5073	20135.07	351.6	
Ar$^+$	330	497.2157	20106.39	380.3	
Ar$^+$	1500	500.9334	19957.16	529.5	
Ar$^+$	620	501.7160	19926.03	560.6	
Ar$^+$	1400	506.2036	19749.39	737.3	
Ar$^+$	10	509.0496	19638.98	847.7	
Ar$^+$	360	514.1790	19443.06	1043.6	
Ar$^+$	1000	514.5319	19429.73	1056.9	0
Ar$^+$	8	516.2745	19364.14	1122.5	65.6
Ar$^+$	38	516.5774	19352.79	1133.9	76.9
Ar$^+$	41	517.6233	19313.69	1173.0	116.0
Ar$^+$	20	521.6816	19163.44	1323.2	266.3
Ar$^+$	150	528.6895	18909.43	1577.2	520.3
Ar$^+$	12	530.5690	18842.45	1644.2	587.3
Ar$^+$	18	539.7522	18521.87	1964.8	907.9
Ar$^+$	11	540.2604	18504.45	1982.2	925.3
Ar$^+$	12	540.7348	18488.21	1998.5	941.5
Ar$^+$	19	545.4307	18329.04	2157.6	1100.7
Ar	14	549.5876	18190.47	2296.2	1239.3
Ar$^+$	14	549.8185	18182.76	2303.9	1247.0
Ar$^+$	14	550.0334	18175.66	2311.0	1254.1
Ar$^+$	22	555.4050	17999.88	2486.8	1429.8
Ar	30	555.87031	17984.81	2501.9	1444.9
Ar	12	557.25248	17940.15	2546.5	1489.6
Ar$^+$	18	557.7689	17923.59	2563.1	1506.1
Ar$^+$	11	557.8518	17920.93	2565.7	1508.8
Ar	48	560.67341	17830.75	2655.9	1599.0
Ar$^+$	14	563.5684	17770.69	2716.0	1659.0
Ar$^+$	14	563.5882	17738.53	2748.1	1691.2
Ar	29	565.07054	17692.00	2794.7	1737.7
Ar$^+$	27	565.4450	17680.28	2806.4	1749.4
Ar$^+$	22	567.2952	17622.62	2864.1	1807.1
Ar$^+$	27	569.1650	17564.73	2921.9	1865.0
Ar$^+$	23	572.4325	17464.47	3022.2	1965.3
Ar	16	573.95207	17418.23	3068.4	2011.5
Ar$^+$	69	577.2326	17319.24	3167.4	2110.5
Ar$^+$	16	578.6560	17276.63	3210.0	2153.1
Ar$^+$	49	581.2746	17198.80	3287.9	2230.9
Ar$^+$	18	584.3781	17107.47	3379.2	2322.3
Ar$^+$	12	587.0443	17029.77	3456.9	2400.0
Ar	11	588.26250	16994.50	3492.2	2435.2
Ar	18	588.85851	16977.31	3509.4	2452.4
Ar	38	591.20861	16909.81	3576.9	2519.9
Ar	10	592.88124	16862.11	3624.6	2567.6

Table 4.1. (continued)

Origin	Relative Intensity	Wavelength (air) (nm)	Wavenumber (vac) (cm^{-1})	Shift (cm^{-1}) relative to 488.0 nm	Shift (cm^{-1}) relative to 514.5 nm
Ar$^+$	11	595.0905	16799.51	3687.2	2630.2
Ar$^+$	23	598.5920	16701.24	3785.4	2728.5
Ar$^+$	20	598.9339	16691.71	3795.0	2738.0
Ar	57	603.21291	16573.30	3913.4	2856.4
Ar	37	604.32254	16542.87	3943.8	2886.9
Ar$^+$	14	604.4468	16539.47	3947.2	2890.3
Ar$^+$	14	604.6894	16532.84	3953.8	2896.9
Ar$^+$	14	604.9072	16526.89	3959.8	2902.8
Ar	15	605.93735	16498.79	3987.9	2930.9
Ar$^+$	11	607.7431	16449.77	4036.9	2980.0
Ar$^+$	91	610.3546	16379.38	4107.3	3050.4
Ar$^+$	1750	611.4929	16348.90	4137.8	3080.8
Ar$^+$	100	612.3368	16326.36	4160.3	3103.4
Ar$^+$	97	613.8660	16285.69	4201.0	3144.0
Ar$^+$	1400	617.2290	16196.96	4289.71	3232.8
Ar$^+$	26	618.7136	16158.10	4328.6	3271.6
Ar$^+$	26	623.9713	16021.95	4464.7	3407.8
Ar$^+$	590	624.3125	16013.19	4473.5	3416.5
Ar$^+$	16	632.4414	15807.37	4679.3	3622.4
Ar	11	638.47189	15658.07	4828.6	3771.7
Ar$^+$	14	639.6614	15628.95	4857.1	3800.8
Ar$^+$	160	639.9215	15622.60	4864.1	3807.1
Ar$^+$	50	641.63075	15580.98	4905.7	3848.8
Ar$^+$	27	643.7604	15529.44	4957.2	3900.3
Ar$^+$	22	644.1908	15519.06	4967.6	3910.7
Ar$^+$	16	644.3858	15514.36	4972.3	3915.4

Table 4.2. (Part One) Principal neon gas lamp lines used for wavelength calibration (taken from CRC Handbook [5])

Ne	Relative Intensity	Wavelength (air) (nm)	Ne	Relative Intensity	Wavelength (air) (nm)
I	15	470.439	I	500	585.249
I	12	470.886	I	100	587.283
I	10	471.007	I	100	588.190
I	10	471.207	I	60	590.246
I	15	471.535	I	60	590.643
I	10	475.273	I	100	594.483
I	12	478.893	I	100	596.547
I	10	479.022	I	100	597.463
I	10	482.734	I	120	597.553
I	10	488.492	I	80	598.791
I	4	500.516	I	100	602.999
I	10	503.775	I	100	607.434
I	10	514.494	I	80	609.616
I	25	533.078	I	60	612.845
I	20	534.109	I	100	614.306

Table 4.2 (continued)

Ne	Relative Intensity	Wavelength (air) (nm)	Ne	Relative Intensity	Wavelength (air) (nm)
I	8	534.328	I	120	616.359
I	60	540.056	I	250	618.215
I	5	556.277	I	150	621.728
I	10	565.666	I	150	626.650
I	5	571.923	I	60	630.479
I	12	574.830	I	100	633.443
I	80	576.442	I	120	638.299
I	12	580.445	I	200	640.225
I	40	582.016	I	150	650.653

Table 4.2 (Part Two)   Principal argon gas lamp lines used for wavelength calibration (taken from the CRC Handbook [5])

Ar	Relative Intensity	Wavelength (air) (nm)	Ar	Relative Intensity	Wavelength (air) (nm)
II	400	465.790	I	20	565.070
I	15	470.232	I	10	573.952
II	20	472.159	I	5	583.426
II	550	472.687	I	10	586.031
II	50	473.205	I	15	588.262
II	300	473.591	I	25	588.858
II	800	476.487	I	50	591.208
II	550	480.602	I	15	592.881
II	150	484.781	I	5	594.267
II	50	486.591	I	7	598.730
II	800	487.986	I	5	599.899
II	70	488.904	I	5	602.515
II	20	490.475	I	70	603.212
II	35	493.321	I	35	604.322
II	200	496.508	I	10	605.272
II	50	500.933	I	20	605.937
II	70	501.716	I	7	609.880
II	70	506.204	I	10	610.564
II	20	509.050	II	100	611.492
II	100	514.178	I	10	614.544
II	70	514.531	I	7	617.017
I	5	515.139	II	150	617.228
I	15	516.228	I	10	617.310
II	25	516.577	I	10	621.250
I	20	518.775	I	5	621.594
II	20	521.681	II	25	624.312
I	7	522.127	I	7	629.687
I	5	542.135	I	15	630.766
I	10	545.165	I	7	636.958
I	25	549.587	I	20	638.472
I	5	550.611	I	70	641.631
I	25	555.870	II	25	648.308
I	10	557.254	I	15	653.811
I	35	560.673			

intensities of the plasma emissions varies with the degree of self-absorption taking place in the cavity as the laser is detuned.

Low pressure discharge lamps are a suitable alternative source of calibration lines. The most useful for Raman spectroscopy, using argon ion or krypton ion laser excitation, are the neon and argon lamps. The emission lines have been accurately measured and are tabulated in Table 4.2. These small lamps are easy to use, readily available and have well documented energy transitions [3–6]. A low pressure mercury lamp also provides useful lines. Hollow cathode discharge lamps can also be used for instrument calibration. Iron, chromium and zirconium are three of the most useful, each having intense emission lines throughout the visible spectrum although they are more bulky than the rare gas lamps and can be more difficult to position in the focal plane of the sample or directly in front of the entrance slits of the spectrometer. Whichever lamp is chosen, it is important to ensure that light enters the spectrometer on-axis. The more intense sources are preferred because a narrower slit-width can be set to ensure that this is the case.

The more emission lines that can be included in a calibration measurement, the better. The best method of calibrating a cosecant driven Raman spectrometer is to perform a simple linear least squares fit of peak position (wave number steps) against absolute line frequencies. This assumes errors only in the measured peak positions. The line of best fit is then used to calibrate the actual frequencies of each data point. If the Raman dataset originates from a spectrograph with a flat focal plane, the same method can be used, this time fitting (channel numbers) against absolute line wavelengths. The following short piece of FORTRAN code illustrates how this can be done:

```
C
C        Perform linear regression for all peaks with assigned
C        wavelengths > 0.0 nm.
C
         SUMW=0.0
         SUMC=0.0
         SUMWC=0.0
         SUMW2=0.0
         SUMC2=0.0
         SAN=0.0
         DO 20 I=1,MAXSP
         IF(HWAVE(I).LE.0.0) GOTO 20
         SAN=SAN+1.0
         SUMW=SUMW+HWAVE(I)
         SUMC=SUMC+DBLE(CNTR(I))
         SUMWC=SUMWC+HWAVE(I)*DBLE(CNTR(I))
         SUMW2=SUMW2+HWAVE(I)**2
         SUMC2=SUMC2+(DBLE(CNTR(I)))**2
   20    CONTINUE
C
         DENOM=SUMC*SUMC-SAN*SUMC2
         DISP=(SUMC*SUMW-SAN*SUMWC)/DENOM
         DOFSET=(SUMC*SUMWC-SUMW*SUMC2)/DENOM
         SIGMAW=SQRT((SUMW2-SUMW*SUMW/SAN)/(SAN-1.0))
         SIGMAC=SQRT((SUMC2-SUMC*SUMC/SAN)/(SAN-1.0))
         R=DISP*SIGMAC/SIGMAW
```

```
C
          DO 30 I=1,MAXSP
          IF(HWAVE(I).LE.0.0) GOTO 30
          SWAVE(I)=(DOFSET+DISP*DBLE(CNTR(I)))
          DWAVE(I)=HWAVE(I)-SWAVE(I)
          SCNTR=(HWAVE(I)-DOFSET)/DISP
          DCHNL(I)=(DBLE(CNTR(I))-SCNTR)
    30    CONTINUE
C
C         Turn the printer on and display the result !
C
          CALL PRINTR(1)
C
          TYPE 650,DISP
          TYPE 660,DOFSET
          TYPE 670,R
          TYPE 680
          TYPE 681
   650    FORMAT(/,5X,' ** Dispersion (nm/chnl.) = ',F10.8)
   660    FORMAT(/,5X,' ** Channel Zero (nm) = ',F9.5)
   670    FORMAT(/,5X,' ** Correlation coefficent = ',F12.10,/)
   680    FORMAT(7X,'Peak',12X,'True ',9X,'Calc.',2(9X,'Diff.'))
   681    FORMAT(6X,'channel',4X,3(4X,'wavelength'),5X'channel',/)
          DO 40 N=1,MAXSP
          IF(HWAVE(N).LE.0.0) GOTO 40
          TYPE 690,CNTR(N),HWAVE(N),SWAVE(N),DWAVE(N),DCHNL(N)
   690    FORMAT(5X,F8.3,8X,F9.4,5X,F9.4,5X,F10.7,4X,F10.6)
    40    CONTINUE
C
```

If, perhaps due to a pin-cushion type distortion, the focal plane of the detector is not flat then a linear least squares calibration cannot be made. The best that can be done is to use an appropriate form of curve fitting. The technique for doing this will not be described here, but references may be found in a later section.

Standard or reference materials are often used by Raman spectroscopists as quick checks of instrument calibration and also tend to be used to compare the performance of different instruments. Solid materials are not really satisfactory for this purpose because internal crystal lattice stresses may shift vibrational frequencies. However, the Raman spectra of sulphur, L cystine, bismuth oxide and mercury chloride are given in Fig. 4.2 for the sake of completeness. The two liquid reference materials most widely used are indene and tetrachloromethane, these are shown in Fig. 4.3. The rotational spectrum of air is shown in Fig. 4.4 and is often used to demonstrate the performance of a monochromator close to the excitation frequency.

4.4 Frequency Response Calibrations

In order to make meaningful comparison of spectra obtained from different Raman instruments it is essential that the frequency response of the collection optics, optical monochromator and detector is normalised. This procedure is also crucial if resonance Raman experiments are used to obtain excitation profiles for one or more Raman

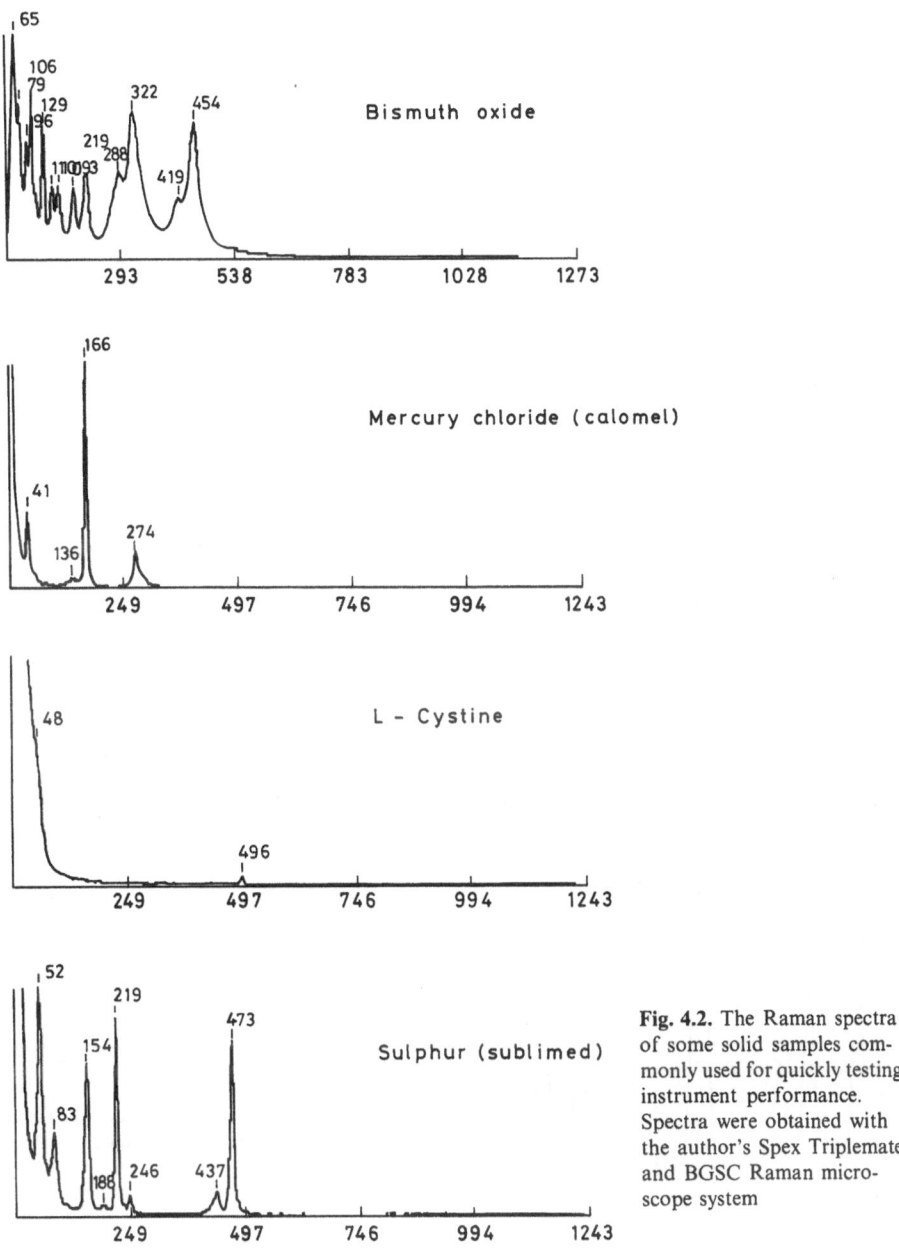

Fig. 4.2. The Raman spectra of some solid samples commonly used for quickly testing instrument performance. Spectra were obtained with the author's Spex Triplemate and BGSC Raman microscope system

modes. The characteristics of dielectric coatings on collection optics are sufficiently good to ignore their effect on spectral throughput. However, the use of beamsplitters in the optical train of Raman microscopes has been noted as an important influence on polarisation measurements [7]. The spectral response of different detectors are

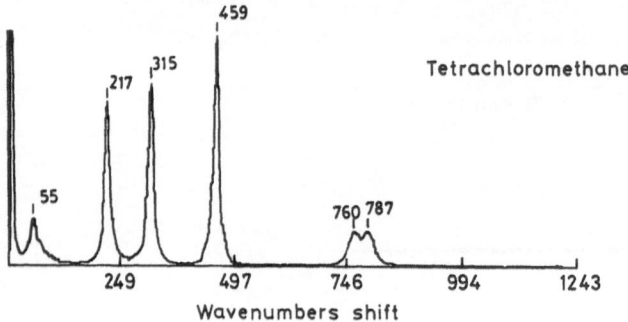

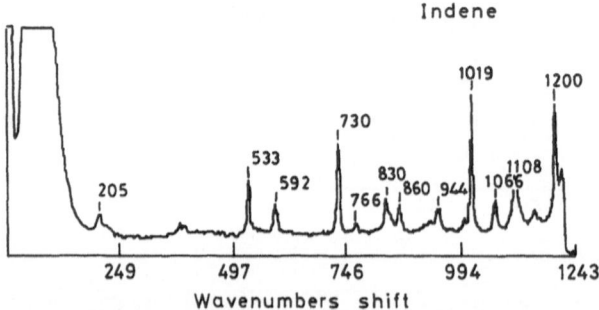

Fig. 4.3. The Raman spectra of indene and tetrachloromethane, two liquids useful for checking spectrograph and spectrometer accuracy and resolution. Results show reasonable accuracy for the author's Triplemate, with a 1200 gpm grating in the final spectrograph stage

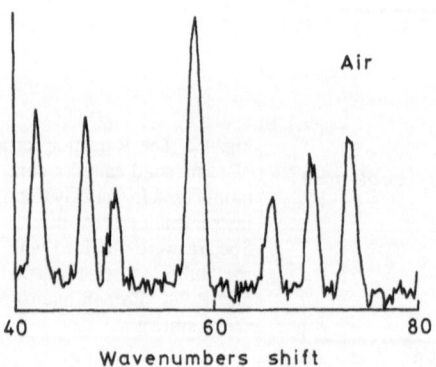

Fig. 4.4. Rotational air spectrum obtained from a Raman microscope system at Newcastle Polytechnic

readily available from manufacturers and have been reproduced here for reference purposes. Quantifying the response of an optical monochromator is rather more difficult. For the purposes of this section, the quantum efficiency of detectors as a function of frequency will be considered first and then the aspects of monochromator corrections will be outlined. In both instances, suggestions are made as to how best apply what information is offered by instrument manufacturers.

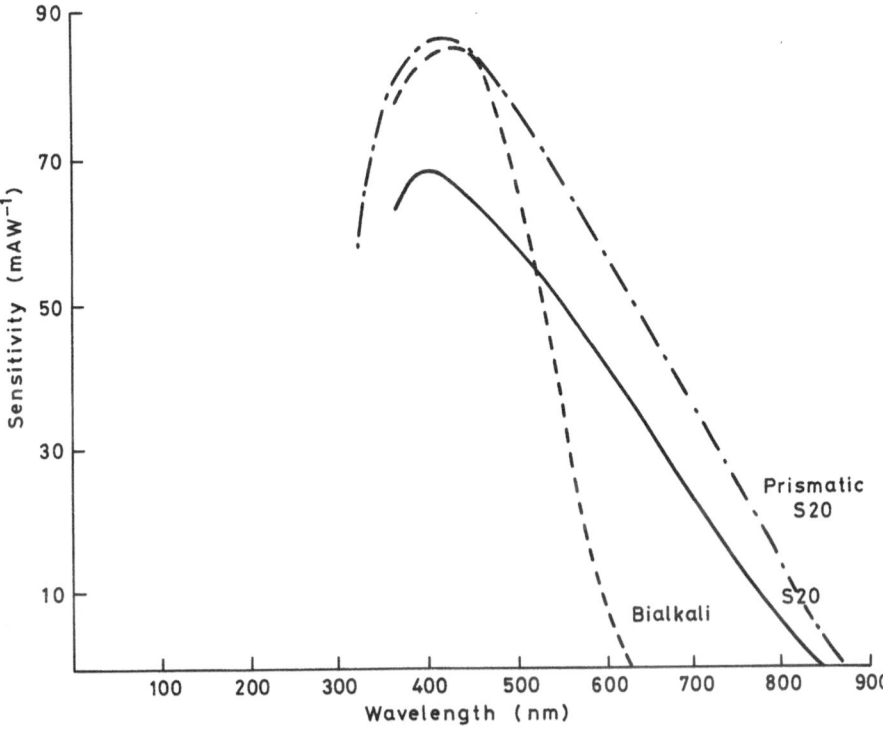

Fig. 4.5. Spectral response curves for photocathode materials used in photomultiplier tubes (courtesy of EMI Ltd.)

Table 4.3 Spectral response values for S20 and Bialkali photocathode materials

Wavelength (nm)	Sensitivity (maW^{-1}) S 20	Bialkali
400	61.4	84.1
450	60.0	84.5
500	54.5	64.1
550	48.6	34.1
600	41.4	7.3
650	32.3	—
700	26.8	—
750	19.5	—

Figure 4.5 has the spectral response curve for each type of photocathode material used in photomultiplier tubes. Unfortunately, this information is not of immediate use. Therefore the responses of an S20 and a bialkali photocathode have been tabulated (Table 4.3). This data can be used to define a polynomial function which defines the frequency response of the detector.

Table 4.4 Spectral irradiance values for a 100 W tungsten halogen lamp operating at 3200 K

Wavelength (nm)	Irradiance ($\mu W\ cm^{-2}$ at 50 cm)
350	0.148
400	0.297
450	0.551
500	0.848
550	1.208
600	1.526
650	1.908
700	2.120
750	2.332
800	2.438

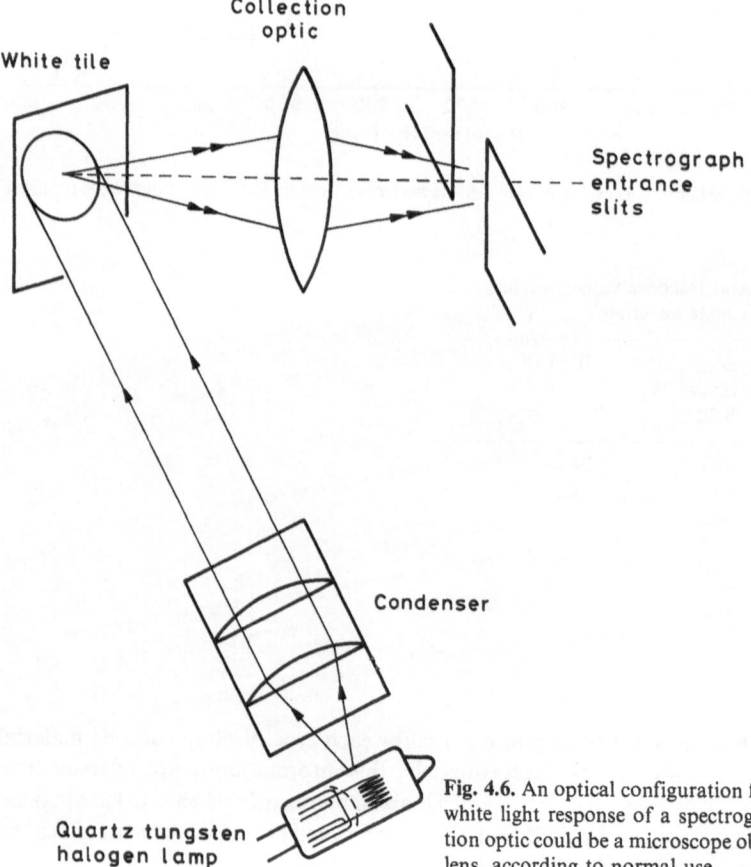

Fig. 4.6. An optical configuration for calibrating the white light response of a spectrograph. The collection optic could be a microscope objective or camera lens, according to normal use

Linear and two-dimensional multichannel detectors that employ microchannel plate intensifiers require careful calibration. Response curves for photodiode array detectors and CCDs can be obtained, but they cannot be used without reference to the spectral response of the photocathode material on the microchannel plate. Because a multichannel detector should really be regarded as an integral part of the spectrograph, it is best to calibrate the optical throughput of the spectrograph and detector together. The method of sample illumination and light collection makes this procedure more or less difficult. Whatever sampling arrangements are chosen, it is important to ensure that a pure white light source of known polarisation (preferably depolarised) is introduced, on axis, into the spectrograph. The best optical configuration for calibrating the white light response of a spectrograph is to reflect collimated white light from a quartz tungsten-halogen lamp, with the aid of a white tile, directly into the collection optics (see Fig. 4.6). This ensures that the coil image of the light source is not reconstructed within the spectrometer, thus achieving a good, even illumination throughout. It is worth remembering that the spectral irradiance of a lamp operating at about 3200 K increases by a factor of approximately 4 times from 400 nm to 700 nm, therefore the 'white light response' of the detector needs to be divided by the spectral irradiance of the lamp. Values are given in Table 4.4. The spectral irradiance curve should be least squares fitted to a good polynomial using the method described in Sect. 4.5.1. Each point of the recorded white light spectrum can then be rationed to the lamp irradiance. This procedure seems cumbersome, but the irradiance function need only by calculated once and the white light response of the spectrograph must be measured at each setting of the centre of the spectral bandpass. In this way the effects of uneven detector illumination, microchannel plate performance and non-uniform detector cooling are perfectly corrected. For the cost of a small amount of software writing, the Raman spectroscopist can and should aim to make meaningful Raman intensity measurements.

4.5 Methods of Data Analysis — Fitting and Smoothing, Adaptive Filtering and Estimation

After dealing with the need to generate Raman datasets whose numerical values are meaningful, it is possible to set about analysis. The problems that beset Raman spectroscopy are well known:

(i) poor signal-to-noise ratios,
(ii) unwanted background signals,
(iii) instrumental line broadening, and
(iv) in some systems, particularly solids, the fact that expected Raman line shapes are complicated and sometimes unknown.

Modern analysis techniques are increasingly used today to solve these difficult problems. Their success usually depends on how much is already known about the scattering process or the source of data corruption (noise). They can be considered under the following headings: smoothing and fitting, adaptive filtering and construction of linear predictors, spectral estimation and spectral synthesis. The first two headings describe techniques that work on the recorded Raman datasets and are distinct from the final group of methods that start with various assumptions about the data content of the Raman spectrum and try to construct a noise-free estimate of the spectrum subject to some constraints. The situation can be thought of in terms of a simple diagram.

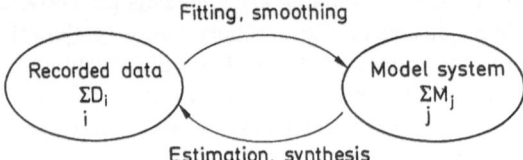

The recorded data and the model are related by the simple expression:

$$\sum_i D_i = \sum_i \sum_j F_{ij} M_j + \sigma_i \tag{1}$$

where F_{ij} is the instrument function and σ_i is the corrupting noise.

4.5.1 Smoothing and Fitting

The first smoothing algorithm to be widely used for all forms of spectroscopy was devised by Golay and Savitsky [7] in 1964. Amended polynomial functions were subsequently published by Steiner et al. [8]. The method operates directly on the data, performing a simple numerical convolution of the data with a window function whose values are those of the coefficients of a cubic or quadratic polynomial. The method does indeed smooth noisy data but, like any window function, it constrains the data to the form of the window. If Raman lines were exactly described by cubic or quadratic functions then the method would work perfectly, providing a least squares fit of the window function to the noisy Raman line shape. However, this is not the case and the window function distorts the line shape to a greater or lesser extent. Even using a good window function, spurious spectral features can be introduced due to the finite width of the convolution function [4]. For this reason, Gans and Gill [10] have suggested the use of spline functions for smoothing and differentiation of spectroscopic data. Spline functions have the advantage over simple window convolutions that although they must be fitted to individual peaks in a spectrum, the number of polynomials and the order of polynomial used for fitting can be chosen. Figure 4.7 illustrates how a simple spline function might work. Each function, f_j, must be evaluated in terms of a polynomial of order k. If the spline has m pieces, it must have m − 1 knots tying them together. Therefore the problem is to evaluate the k + m number of coefficients of the polynomials f_j, so as to meet a least squares criterion with respect to the data. The methods by which this can be done will not be described here, but FORTRAN versions of cubic spline fitting codes (E02BAF and E02BCF) can be obtained from the NAG library [11], the Harwell Subroutine Library [12] or de Boor [13]. Spline fitting can also be used, with a little intuitive help

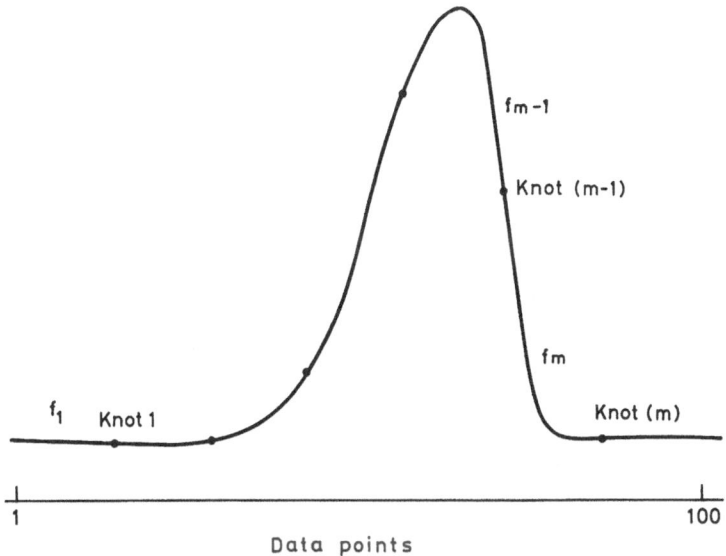

Fig. 4.7. An illustration of how a simple spline function with 6 knots and 7 pieces might be used to fit an experimental lineshape

from the spectroscopist, to remove slowly varying background features. Such features may be due to the elastic scattering background or sample fluorescence. Ten or twenty points can be chosen, well away from spectral lines, as representative of the background. These data values are then used to evaluate the polynomial coefficients of the spline, from which the whole background contribution can be simply interpolated. This method of spline fitting can also be used to construct calibration curves for detector response and to normalise the instrumental throughput of spectrographs.

4.5.2 Adaptive Filters and Adaptive Smoothing

The methods discussed so far have made no use of basic noise statistics that may be known a priori to data processing. Techniques that make use of this knowledge or glean it form the actual data are termed 'adaptive'. Hence we have two popular methods, adaptive smoothing and adaptive filtering. It is not appropriate to provide a detailed description of these techniques in this text or to give a formal introduction to statistical analysis. Two excellent texts on this subject are 'Statistics for Technology', (pub. Chapman and Hall, Third Edition, 1983, London and New York), and 'The Analysis of Time Series', (pub. Chapman and Hill, Third Edition, 1984, London and New York), both written by C. Chatfield. Raman datasets can basically be viewed in two ways for statistical purposes. Either they represent a number of measured variables, related by their covariances to one another or they can be thought of as a single measured variable (photon count rate for example) that varies with time, spectral features being related to the autocorrelation function of that variable. Therefore we have, crudely speaking, the two approaches of data fitting and time series analysis. Least squares fitting a cubic spline function is an example of data fitting and adaptive smoothing is an example of time series analysis.

The first adaptive smoothing technique that has been useful for improving the signal-to-noise ratio of Raman spectra was developed by Kawata and Minami [14] and demonstrated by Tanabe and Hiraishi [15]. The method of Kawata and Minami is significantly better than Savitsky-Golay smoothing, it produces markedly less spurious 'peaks' which may be attributed to low frequency components of the noise and tends to retain the original line shape of Raman bands. Unlike Savitsky-Golay, Kawata-Minami smoothing attempts to find the spectral estimate $\hat{M}_j$ such that Eq. (2) approximates to the description of a real data set given in Eq. (1)

$$\sum_i D_i = \sum_j \hat{M}_j + \sum_i \sigma_i \tag{2}$$

where

$$\sum_j \hat{M}_j = \sum_{ij} F_{ij} M_j \tag{3}$$

F_{ij} therefore is the instrument function and is approximated by the weighting function used in either form of smoothing. It is important to realise why the approach of Savitsky and Golay is inadequate. It assumes that the statistics of the noise σ_i is uniform across the spectrum. This is clearly not true. When the photon count rate is low, the noise will be dominated by detector dark current and thermal noise from instrumentation, however, at high photon count rates the noise will be dominated by photon counting statistics. Realising this, Kawata and Minami normalise the effect of the smoothing function so that

$$\sum_j \hat{M}_j = \sum_i a_i \cdot [X_i - \bar{X}_i] + \bar{X}_i \tag{4}$$

where $\bar{x}_i$ is again the dataset, $\bar{x}_i$ represents the local mean (over a number of arbitrary points) and a_i is the smoothing coefficient spectrum. a_i is defined as:

$$\sum_i a_i = \sum_i [\sigma_{D_i}^2 - \sigma_n^2]/\sigma_{D_i}^2 \tag{5}$$

where $\sigma_{D_i}^2$ is the local variance and σ_n^2 is the 'prior knowledge' noise variance of the detector and counting electronics. A simple FORTRAN or BASIC program can easily be written to perform the calculations according to the following algorithm:

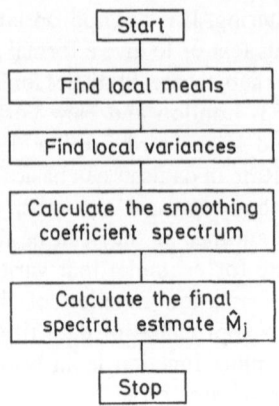

(A)

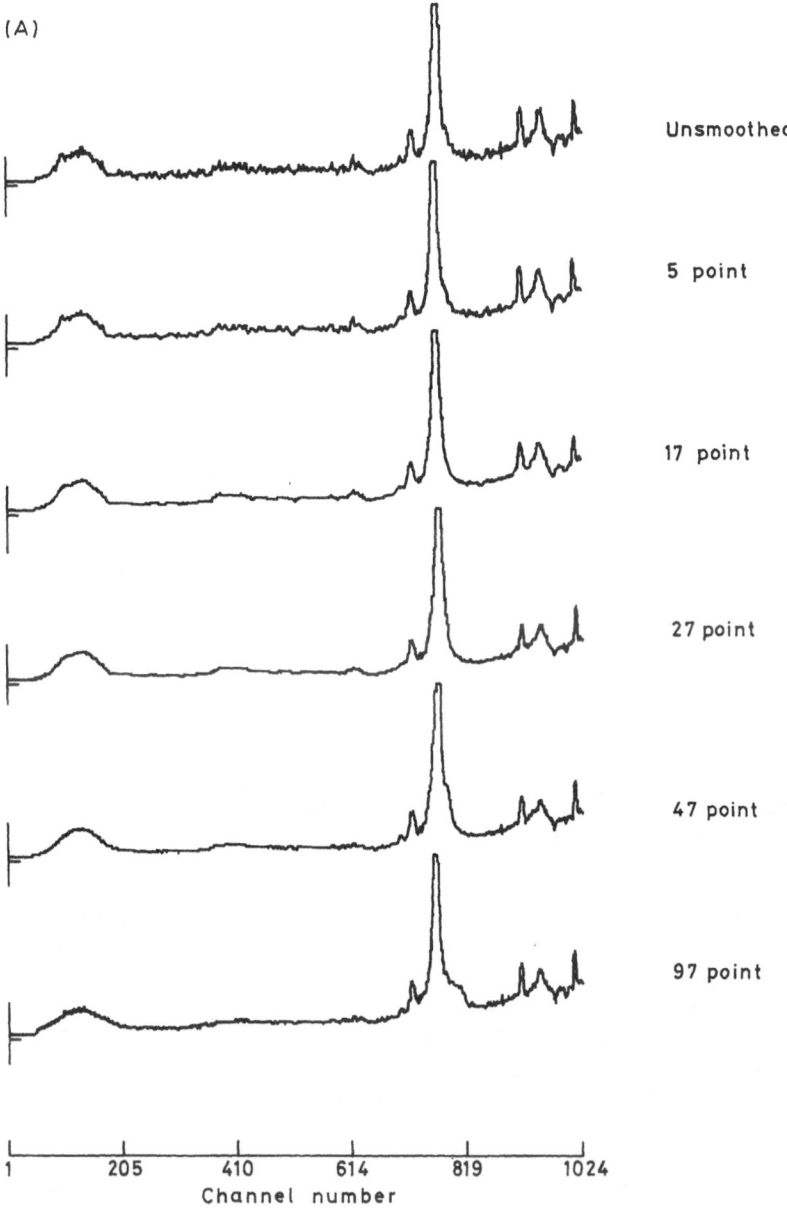

Unsmoothed

5 point

17 point

27 point

47 point

97 point

1   205   410   614   819   1024

Channel number

Fig. 4.8. Kawata and Minami adaptive smoothing has been used to improve the appearance of (a) a chromia Raman spectrum (b) a ferrite Raman spectrum. Any number of points up to the narrowest line's FWHM may be safely used, however spurious shoulders start to appear when the adaptive average exceeds this limit

(B)

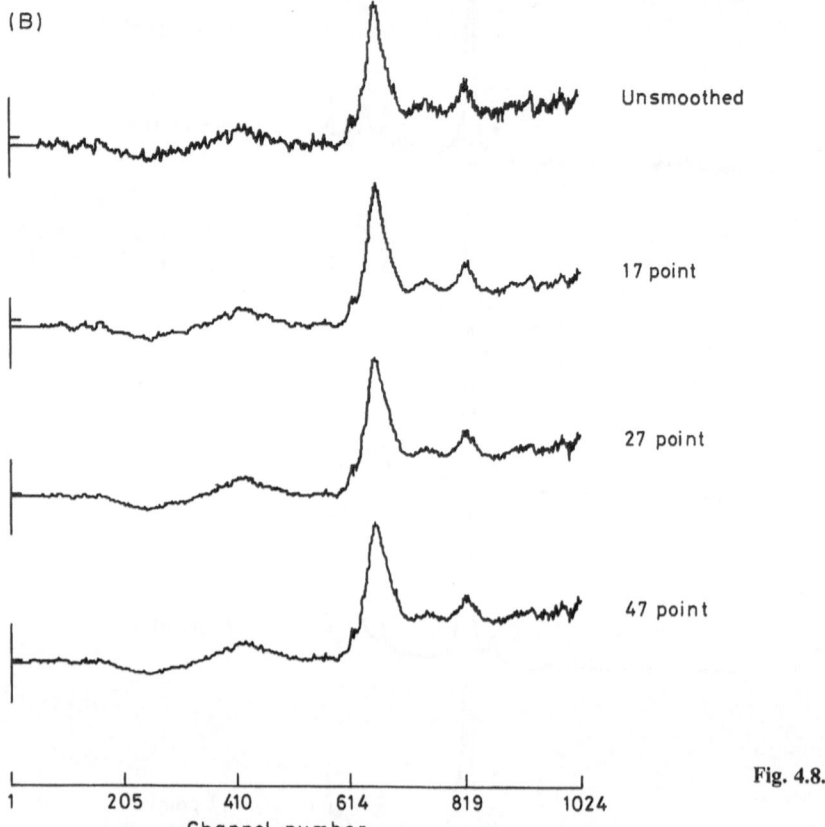

Unsmoothed

17 point

27 point

47 point

Fig. 4.8.

1   205   410   614   819   1024

Channel number

The results of such an algorithm are shown in Fig. 4.8. Figure 4.8a shows an improvement in the apparance of the data up to a 17 point window, which corresponds to the FWHM of the bands. Increasing the window generates a spurious shoulder on the most intense feature. Figure 4.8b indicates that much larger window averages can be used if the Raman bands are broader.

Although the adaptive smoothing algorithm described above is successful in the example of high signal-to-noise spectra, no smoothing technique seems to be successful at dealing with poor signal-to-noise ratio Raman spectra. Since inelastic light scattering is a low probability event, Raman spectra frequently are found to have very poor signal-to-noise ratios. Despite the advantages of modern array-type detectors in signal averaging, spectra of thin solid films, adsorbed monolayers of gases on metal single crystals and catalysts are often ambiguous. In these cases, a robust adaptive filtering technique is the only method reported to date that has proved successful. Dyer and Hardin [16, 17] have proposed two such methods. The first is adaptive peak detection (APD) and the second is matched filtering.

Adaptive peak detection is a two stage process that requires no detailed knowledge of the signal or the exact noise statistics. This makes it ideal for practical purposes. Initially an adaptive filter is used to locate local variations in the noise statistics by using previous values of the data to estimate the expected value of the input signal plus noise at a particular time. The weighting coefficients are governed by the variance of the output of the APD and calculated using an appropriate algorithm [18]. After conditioning the APD, a variance estimator is constructed in the form of a simple moving average. The output of the variance estimator is a delta function that indicates the position of a peak buried within the noise. The APD algorithm, therefore, gives

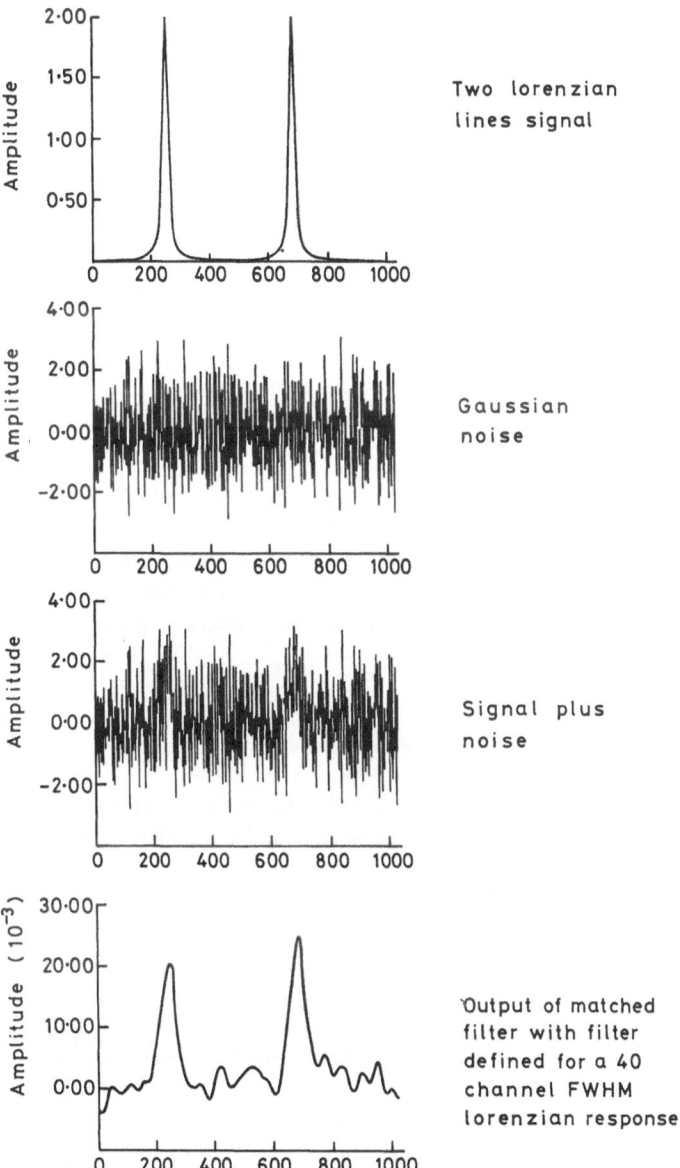

Fig. 4.9. Application of matched filtering to simulated Raman data (After Dyer and Hardin)

no quantitative measure of peak intensity or Raman line shape. It is expected to work best when a Raman band is present in a noisy background with a signal-to-noise ratio of less than 1.0. This algorithm will work well for photon count data recorded with a photomultiplier tube and a slowly scanned spectrometer, but it has not been demonstrated successfully for noisy datasets from a photodiode array detector or CCD. Adaptive peak detection would be expected to perform less well with data from an intensified photodiode array. This is because the noise is not white due to cross-talk, charge leakage from the diodes and microchannel plate blooming. The noise is also not stationary because charge is shared between adjacent diodes. For these reasons, it is suggested that Raman datasets obtained from array detectors are best improved with an adaptive smoothing technique or a spectral estimation method, which is described in the following section.

If the line shape of the expected Raman band or bands is known and the noise corrupting the spectrum is white and stationary, then matched filtering can be used to improve the signal-to-noise ratio. The method has a similar basis to adaptive peak detection. Dyer and Hardin [17] have described a filter that assumes a pure uncorrupted signal, S(E), with some noise component, (E). The purpose of the filter is to maximise the signal to noise ratio of its output, so that

$$\frac{\partial \bar{S}_0}{\partial \bar{N}_0} = 0 \tag{6}$$

where $\bar{S}_0$ is the average output signal power and $\bar{N}_0$ is the average output noise power. They use a Fast Fourier Transform technique to accomplish this which will not be described here. The result of matched filtering simulated Raman data is given in Fig. 4.9 and shows that with careful choice of filter parameters, a much more confident assessment of weak bands can be made. It remains to be seen whether this technique works well for real Raman data. Further reading on the subject of matched filters can be found elsewhere [19].

4.5.3 Spectral Estimation and Spectral Synthesis

All the techniques described thus far have been found to be ineffective at low signal-to-noise ratios or to require some assumptions: the noise and the signal are orthogonal; the noise is white and stationary; the datasets contain no underlying trends. However, there are *image processing* techniques that do not require assumptions or prior knowledge to estimate successfully the spectral or 'real' data from noise or other corrupting effects. These techniques are based on what is called a Maximum Entropy Method (MEM). In fields apart from Raman spectroscopy, maximum entropy is being widely used as a powerful method of image reconstruction [20–23]. The algorithms used in MEM image reconstruction have been published [24, 25] and will not be described in detail here.

The most important distinction between MEM and smoothing or fitting techniques is that the spectral estimate is contructed without manipulation of the original dataset, only applying a statistical check of the estimate against the variance of the original dataset. If a Raman spectrum is considered to be a simple one-dimensional image, corrupted by noise and some extraneous background luminescence, then the objective of an image reconstruction technique is to find the estimate which best fits the observed data, subject to some criteria. Clearly, the number of images that can be constructed form 1024 data points within an instrument's dynamic range (DR) will be given by:

$$(1024) \times DR + 1 . \tag{7}$$

Since this is a very large number, the criteria must be carefully chosen. The most obvious criterion is that the estimate must fit the data to within the noise variation, i.e. that it satisfies a simple minimal χ^2 test

$$\chi^2 \simeq \sum_{i=1}^{1024} \sum_{j=1}^{1024} \frac{1}{1024} \sigma_i^2 [(f_j + b_j) h_{ij} - d_i]^2$$

where σ_i^2 is the data variance, f_j and b_j are spectral and background estimates, d_i is the Raman data and h_{ij} is the spectrometer instrument function. The second criterion is that the entropy, S_f, content of the spectral estimate is maximised. Where S_f is given by

$$S_f = -\sum_j f_j \ln \left(\frac{f_j}{n_j} \right) \tag{8}$$

and likewise the background entropy content S_b is

$$S_b = -\sum_j b_j \ln \left(\frac{b_j}{n_j} \right) \tag{9}$$

Much has been written about the maximal entropy criterion, but perhaps the best proof of its validity is that it works. It is beyond the scope of this book to describe in detail how MEM algorithms are implemented, but details are given by Skilling [24]. The Fourier reconstructions work in the following way:

Step 1 — Select the dataset and instrument or point spread function.
Step 2 — Set the default values for f_{ij} and b_{j2}.
Step 3 — Insert the defaults and start the first reconstructions.
Step 4 — Fourier transform and calculate χ^2, S_f and S_b the entropies, ∇Q the gradients and $\nabla\nabla Q$ the curvatures and finally ∇S the entropy slope.
Step 5 — From the values for gradients and curvatures, calculate the vector required to progress to a minimum χ^2 and to not stray too far from the line of steepest approach on the entropy surface.
Step 6 — With this information, predict the expected value for χ^2 for a small step up the entropy surface.
Step 7 — Check to see if $\nabla S = 0$ and χ^2 is the required minimal value. If not, return to step 3.

The algorithm converges after about 21 iterations, to yield a unique solution. An example of its use is given in Fig. 4.10. The point spread function (PSF) was recorded by measuring the instrumental contribution to the linewidth of a weak

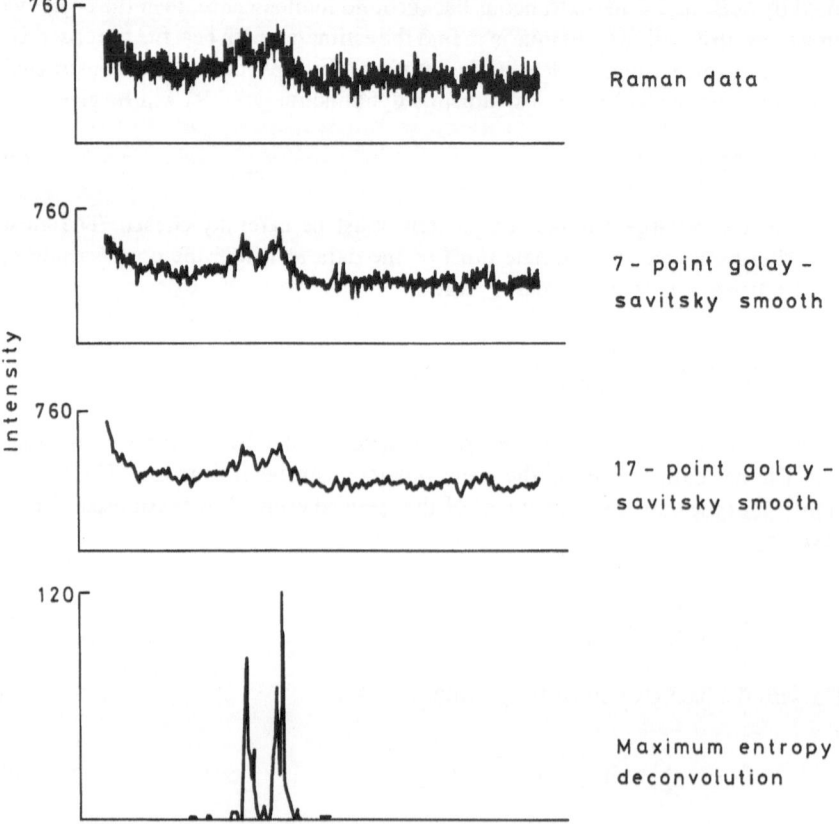

Fig. 4.10. An example of the use of MEM spectral estimation to resolve a poor signal-to-noise ratio Raman spectrum of a carbide

laser plasma line. The program was run on a PDP 11/73 with a floating point co-processor and took approximately 15 minutes. Careful evaluation of the MEM algorithm for Raman spectroscopy has shown [26] that in most cases results are quantitative and excellent separation of signal, background and noise is achieved. The only prior information required is the point spread function, which is used as a Fourier convolution tool for the synthesized estimate. Another example (Fig. 4.1) shows that Raman lines of different half widths are accurately reconstructed. This is of course, because the instrument function is measured and consequently removed from the final spectral estimate.

Smoothing, fitting or filtering are three methods of improving Raman spectra. Very often it is found that specific questions need to be answered on the basis of spectral evidence such as, "How much of phase A is present in a mixture of phase A plus phase B?", or, "What evidence is there for a phase other than A or B?" In trivial cases, a molecule or crystal's vibrational energies and scattering efficiencies are sufficient to unambiguously finger-print a given component of some mixture. However,

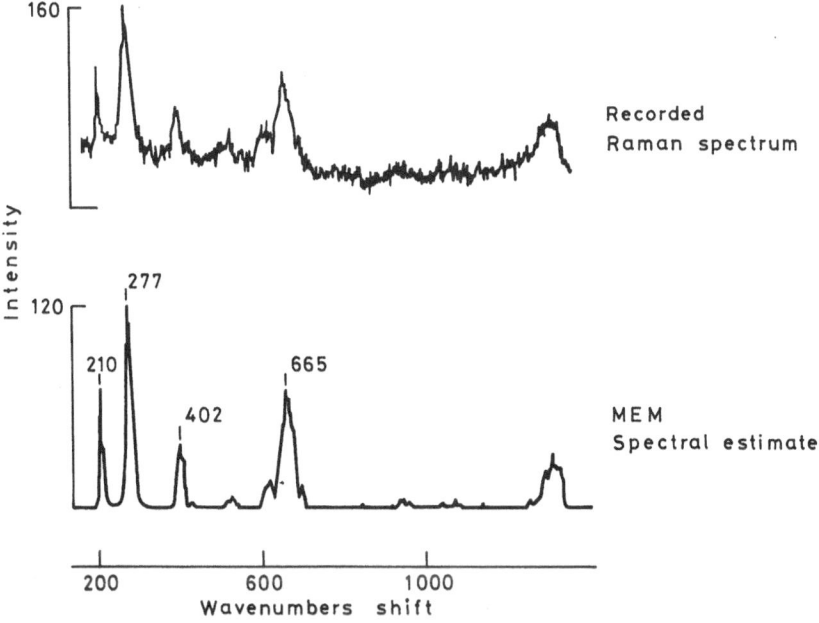

Fig. 4.11. An MEM spectral estimate of the Raman spectrum of Fe_3O_4. Comparison with the original spectrum shows that the line widths of the broad phonon mode at 666 cm^{-1} and the plasma line at 206 cm^{-1} are correctly reproduced

when Raman bands are broad and overlapping, quantitative measurements can be difficult. If an unknown phase is present in a mixture, some test must be applied. A straightforward mathematical technique exists to solve both these problems, and provides a measure of how accurate a particular concentration measurement is

$$\sum_i d_i = \sum_i ax_i + bx_i + \ldots + nx_i + \sum_i \sigma_i . \tag{10}$$

Equation (10) describes a Raman dataset d_i composed of the Raman spectra of n phases with a random error σ_i associated with the measurement. Assuming the error to be Gaussian white noise, it is possible to find, by a method of least squares, values for the weighting coefficients a, b, ... , n if reference Raman datasets of the n phases are available. The FORTRAN subroutine MA44, available from the Harwell Subroutine Library [12], efficiently solves this problem. A similar NAG routine is also available [11]. MA44 works on a linear least squares basis, providing both solution standard deviations and the variance-covariance matrix. Figure 4.12 illustrates how a simple FORTRAN program incorporating MA44 can solve a four component mixture problem.

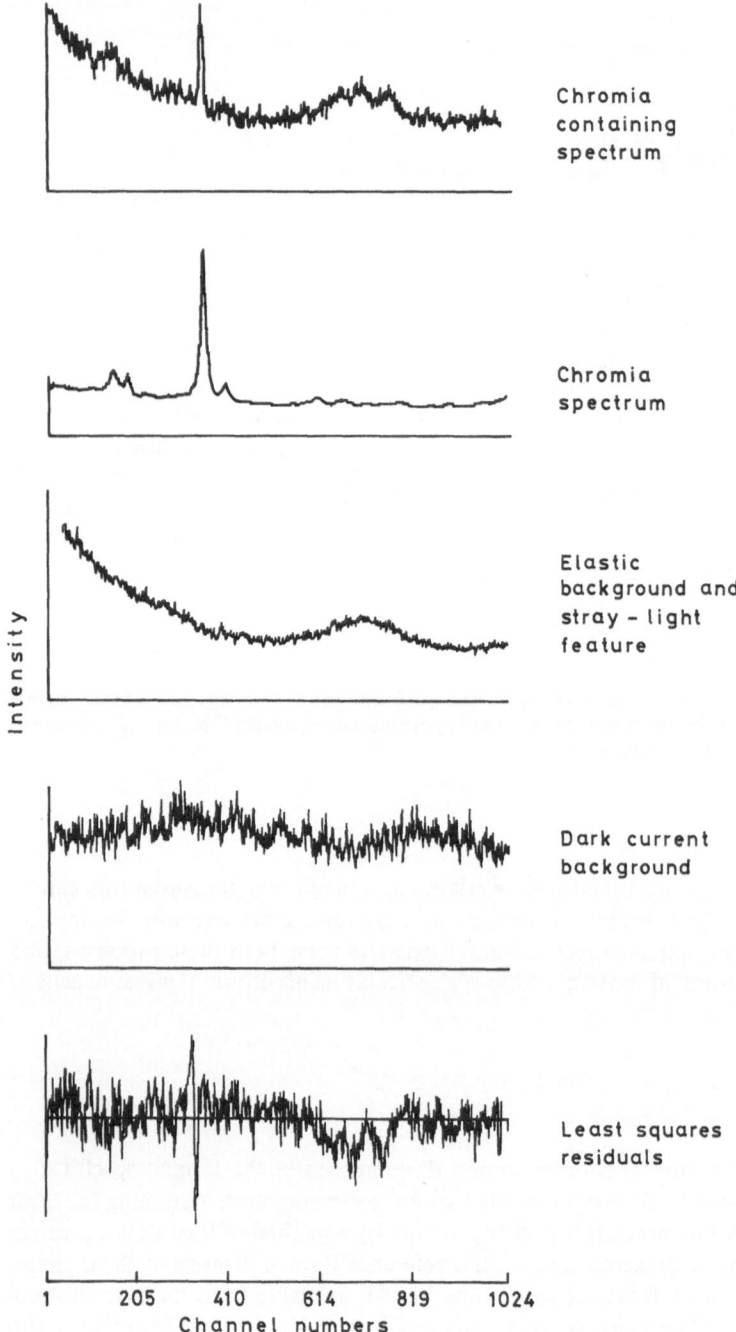

Chromia
containing
spectrum

Chromia
spectrum

Elastic
background and
stray – light
feature

Dark current
background

Least squares
residuals

Fig. 4.12. A demonstration of the use of spectral synthesis of a four component dataset, Rayleigh scatter is rejected and reveals a phase in addition to chromia. The linear least squares Harwell Fortran subroutine MA44 was used to accomplish this

4.6 References

1. Loader J (1970) Basic laser Raman spectroscopy. Heyden, London, chap 4
2. Craig NC, Levin IW (1979) Appl. Spectrosc. 33: 475
3. Striganov AR, Sventitskii NS (1968) Tables of spectral lines of neutral and ionised atoms. IFI/ Plenum Data Corp., New York
4. Savoie R, Pigeon-Gosselin M (1983) Can. J. Spectrosc. 28: 133
5. Chemical Rubber Company Handbook of Chemistry and Physics (1987) 68th edn, Weast RC (ed) CRC, Boca Raton, FL, pp E201–E327
6. Meggers WF, Corliss CH, Schribrer BF (1961) Tables of spectral-line intensities Part 1, NBS Monograph 32. US Government Printing Office Washington 25 DC
7. Savitsky A, Golay MJE (1964) Anal. Chem. 36: 1627
8. Steinier J, Termonia Y, Deltour (1972) Anal. Chem. 44: 1906
9. Gans P, Gill GB (1983) Appl. Spectrosc. 37: 515
10. Gans P, Gill GB (1984) Appl. Spectrosc. 38: 370
11. Numerical Algorithms Group (NAG), 7 Banbury Rd., Oxford OX2 6NN, England
12. Computer Science and Systems Div. Harwell Subroutine Library catalogue AERE Report AERE R9185, Harwell Laboratory England
13. de Boor C (1978) A practical guide to splines. Springer, Berlin Heidelberg New York
14. Kawata S, Minami S (1984) Appl. Spectrosc. 38: 49
15. Tanabe K, Hiraishi J (1984) Computer Enhanced Spectrosc. 2: 17
16. Dyer SA, Hardin DS (1984) J. Raman Spectrosc. 15: 401
17. Dyer SA, Hardin DS (1985) Appl. Spectrosc. 39: 655
18. Widrow B, Hoff ME (1960) 1960 IRE WESCON Conf. Record pt 4, pp 96–104
19. Papoulis A (1984) Probability, random variables and stochastic processes, 2nd edn. McGraw-Hill, New York, p 298
20. Gull SF, Daniell GF (1978) Nature 272: 686
21. Skilling J, Strong AW, Bennett K (1979) Mon. Not. R. astr. Soc. 187: 145
22. Minerbo G (1979) Comp. Graph. Im. Proc. 10: 48
23. Bryan RK, Bansal M, Folkard W, Nave C, Marvin DA (1980) Proc. Nat. Acad. Sci. USA 80: 4728
24. Skilling J (1981) Workshop on maximum entropy estimation and data analysis, Univ. Wyoming, Reidel, Dortrecht, Holland
25. Skilling J, Bryan RK (1984) Mon. Not. R. astr. Soc. 211: 111
26. Graves PR, Gull SF to be published

CHAPTER 5

Non-Standard Physical and Chemical Environments

by Derek J. Gardiner
Raman Applications and Instrument Development Group, Department of Chemical and Life Sciences,
Newcastle upon Tyne Polytechnic, Newcastle upon Tyne, UK, NE1 8ST.

5.1 High Temperature

The facility to measure Raman spectra at variable high temperature is an essential requirement for many areas of study. Detection of solid state structural modifications and reactions, molecular dynamics, conformational equilibria, and species identification and structural characteristics in melts, are amongst the many examples of Raman studies at high temperature. For some room temperature liquids an increase in temperature may often require a sample cell capable of retaining high pressures; this aspect is dealt with later (Sect. 5.3). Here we will look at the problems of attaining moderately high temperature spectra where there is no excessive high pressure requirement.

Most commonly the sample, held in a Pyrex glass or quartz tube, is heated resistively either by winding the heating coils around the tube or by inserting the tube into a heated block. Various degrees of thermal insulation are included depending upon the maximum temperature and the thermal stability required. Temperature is generally monitored by a thermocouple placed close to, or preferably in contact with, the sample. Stability is achieved either by thermal equilibrium, which in some cell designs can take up to 1 to 2 hours for high temperatures to be realised, or by an electrical feedback heating and, or, cooling system.

Formation of bubbles in the focused laser beam causes increased noise and spikes in the spectra and in some sample geometries can be very difficult to eliminate. Where the laser is focused horizontally into the sample, any bubbles formed will be displaced upwards out of the beam. In the case of vertical laser illumination, bubbles can collect in the focused beam at the top of the sample volume. In this case, special care, either in sample preparation, or in the provision of a way for the bubbles to escape, must be taken. In many cases, the problem is simply dissolved air which nucleates into bubbles at the hot spot formed where the focused beam enters the sample or on fine solid

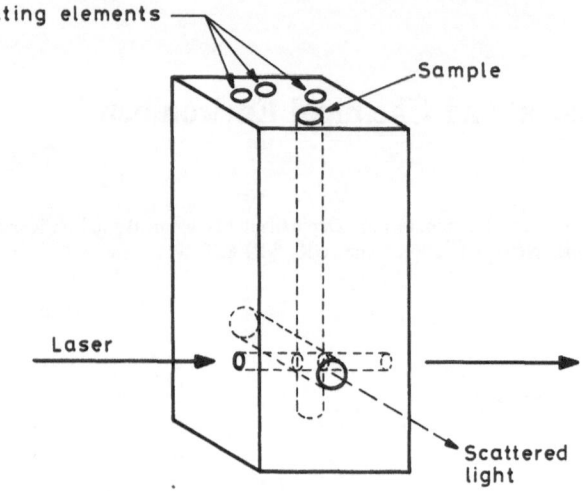

Fig. 5.1. Simple, high temperature Raman cell. The sample is held in a sealed glass tube and the cell block can be mounted on a thermally insulating base, either with horizontal illumination (as shown), or on its side with vertical illumination

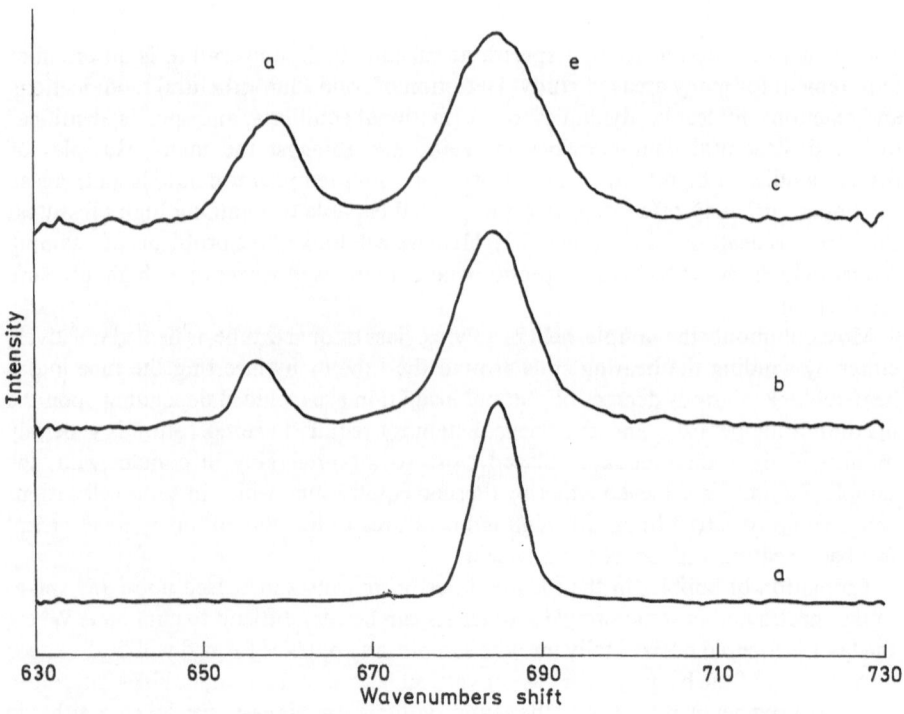

Fig. 5.2. Raman spectra (a, b, c) of chlorocyclohexane at −40, 22 and 110°C. Bands arising from the equatorial and axial conformers are marked e and a respectively. Spectral bandpass approx. 4 cm^{-1} with 300 mW of 514.5 nm laser power at the sample. (Reproduced from reference 2 with permission)

impurities in the sample. Alternatively, it is caused by sample decomposition. In each case, slight defocusing of the laser beam will often improve matters. It is good practice, however, to remove as much of the dissolved air in the sample as possible before recording spectra. This is easily achieved by repeated freezing and pumping on the sample prior to sealing it in the sample tube. In the case of molton salt spectra, pumping just above the melting point is generally effective.

The cell shown in Fig. 5.1 has been used for temperatures up to 200 °C with both horizontal and vertical illumination [1]. The heating block is made from brass, heat is supplied by three cylindrical heating cartridges. In addition the laser beam can be refocused by a mirror back into the sample to increase the signal. A thermocouple monitors the temperature and stability is achieved through equilibration. Using this cell the spectra of the $\nu(C-Cl)$ stretching region of chlorocyclohexane shown in Fig. 5.2 were obtained. The band at 685 cm^{-1} is due to the axial conformer whilst that at 731 cm^{-1} arises from the equatorial conformer. As the temperature is raised from -40 °C to 110 °C the axial band is seen to increase with respect to the equatorial band. Analysis of these data leads to a ΔH^0 value for the conformational change of 1.10 kJ mol^{-1}, in very good agreement with the results of alternative methods [2].

At the other extreme [3], the vacuum furnace shown in Fig. 5.3 has a high degree of sophistication. An externally heated hot-finger element on which the sample is mounted, is inserted into the quartz tube and temperatures up to 1450 K can be achieved. The temperature range can be extended to 1650 K by using an internal platinum coil element. A low current power source is adequate and temperature control of ± 1 K is reported.

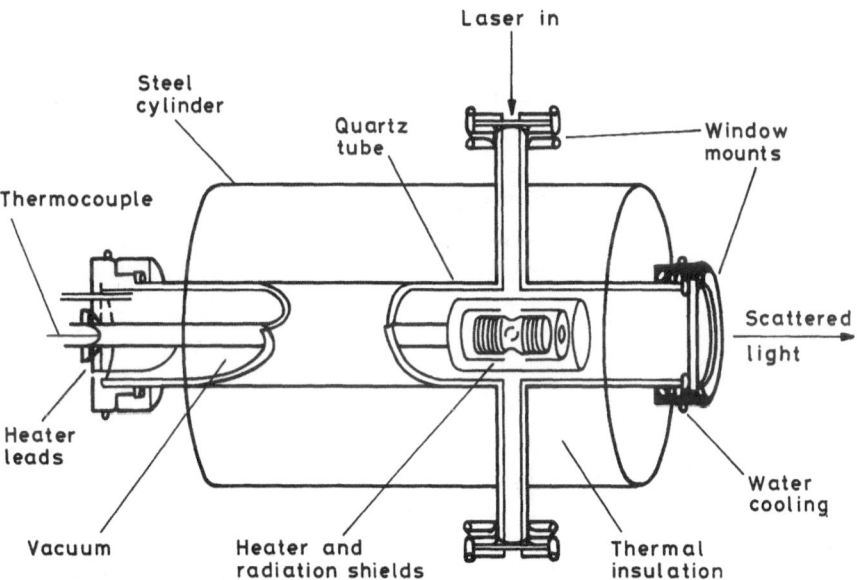

Fig. 5.3. Principal components of a vacuum furnace used to measure Raman spectra at temperatures up to 1375 °C. (Reproduced from reference 3 with permission)

It is also possible to use windowless cells relying on surface tension, to study molten salts like fluorides which may otherwise react with the window material [4, 5].

5.2 Low Temperature and Matrix Isolation

The ability to record Raman spectra from samples at low temperature is an essential in many fields. Solid phase and single crystal work is often much simplified if contributions from hot bands, especially in the lattice mode region, can be removed. Low temperatures are also necessary to maintain the liquid phase for low boiling point

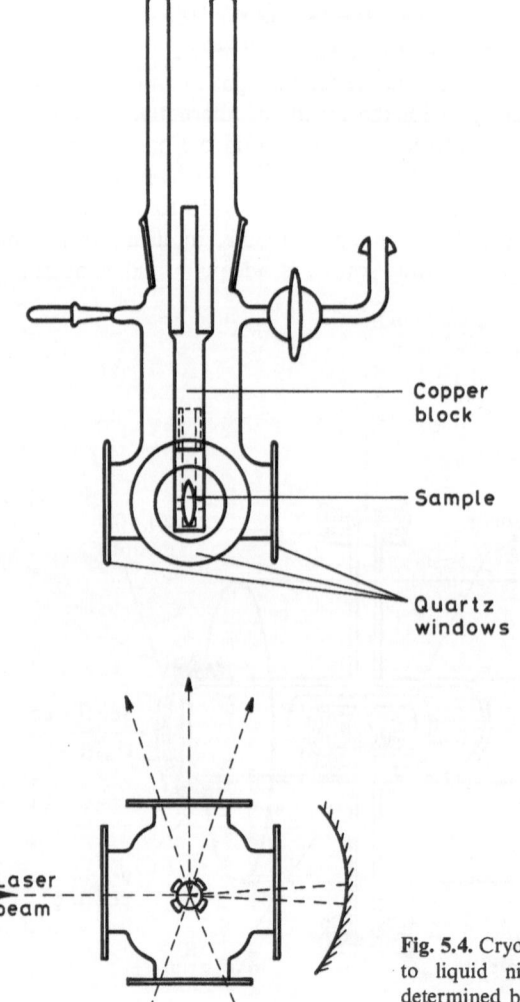

Copper block

Sample

Quartz windows

Laser beam

Fig. 5.4. Cryostat for measuring Raman spectra down to liquid nitrogen temperatures. Temperature is determined by the coolant used and both horizontal and vertical illumination geometries can be accommodated

liquids and extension of the temperature range over which conformational and molecular dynamic studies on liquids can be made leads to much improved data.

A simple vacuum cryostat in which the sample is held in a sealed tube within a cooled copper block is shown in Fig. 5.4. The cryostat has four glass windows for use with a horizontal laser beam geometry and a further window set in the base of the cell allows a vertical laser beam geometry to be used. Attachment of a thermocouple and a heating coil with feedback control would allow regulation of the temperature. A cell of similar design but based upon a low temperature still has also been used [6]. The use of a vacuum jacket can be avoided by heating the windows of a cold cell to stop condensation [7], this also reduces the number of window surfaces that the laser beam and scattered light have to pass through, thus improving the Raman signal.

Matrix isolation allows Raman spectra to be obtained from highly reactive and otherwise unstable species like free radicals, photochemically produced molecular fragments, unusual molecules and reaction intermediates. The technique has been comprehensively reviewed [8, 9]. Low temperature matrix isolation relies on trapping the species to be studied in a surrounding host matrix of a solid inert material. The matrix and solute are condensed onto a cold block made of a copper, silver, or platinum mirror or of a sapphire disk and maintained under high vacuum. Matrix materials are typically inert gases, mostly argon and nitrogen. However, where appropriate, other materials like carbon monoxide, carbon dioxide, water, methane and carbontetrachloride have also been used. The temperatures most often employed are 4 K 20 K and 77 K being the boiling points of helium, hydrogen and nitrogen respectively. Such low temperatures require special cryostats and rely either, in the case of hydrogen, on a constant supply of coolant gas cooled below its inversion temperature and then liquified and delivered to the sample block or for helium and

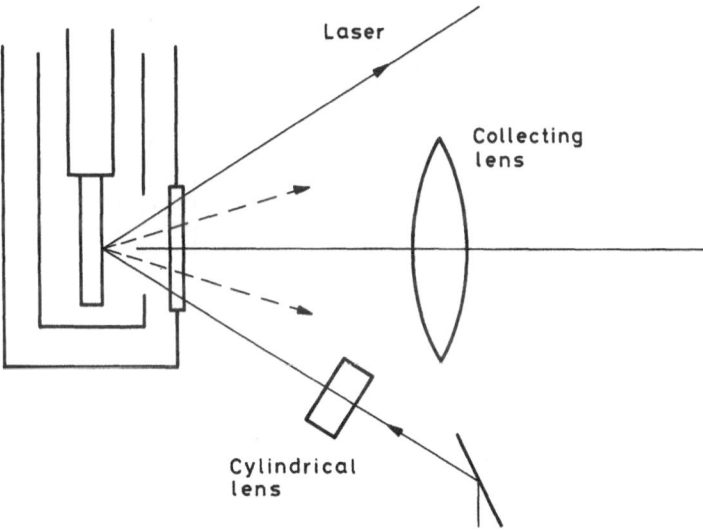

Fig. 5.5. Experimental arrangement for observing matrix isolation Raman spectra using a cylindrical lens to produce a line focus on the sample. (adapted from Ref. 10)

neon, continuous liquid flow cryostats are used. Further temperature variation may be obtained by local heating of the sample block or by lowering the pressure over the coolant. The sample to be studied, its photochemical precursor, or the reactants are mixed with the matrix gas at low concentration, to ensure isolation and then introduced into the evacuated cryostat to condense onto the cold sample block forming a sample film.

A favoured [10] sample block-illumination geometry is shown in Fig. 5.5 which uses a cylindrical lens to produce a line focus of the laser beam on the sample. An alternative method uses a caesium iodide plate as a cold surface on which the matrix is deposited. This arrangement which is shown in Fig. 5.6 in the spray-on geometry, is reported [8] to give comparable results to those obtained from metal surfaces and has the added advantage that the same matrix can also be studied by infrared spectroscopy.

Raman matrix isolation poses particular advantages and disadvantages in comparison with the more widely used infrared technique. Thin films can be used ~10 mi-

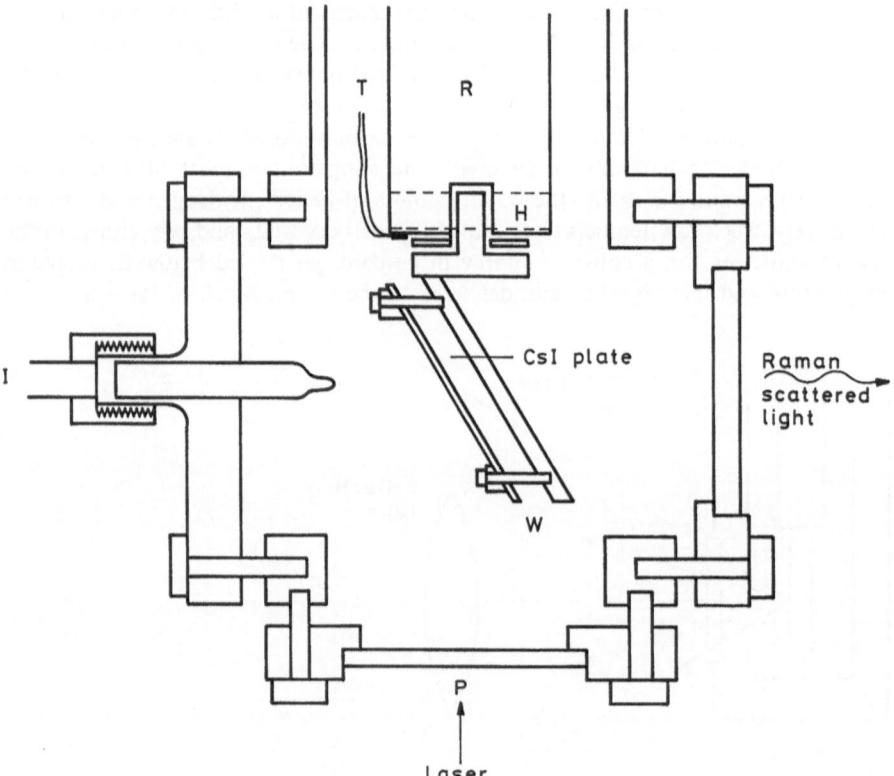

Fig. 5.6. Arrangement used for matrix isolation Raman spectroscopy, in which the sample is pulsed onto the cold surface. The sample deposition configuration is shown. In order to observe Raman spectra the sample plate is rotated through 180° to allow the laser beam to fall on the deposited matrix and for the scattered light to be collected as shown. R = cold station of refrigerator, T = thermocouple, H = heater, p = Pyrex or quartz windows, I = spray-on tube, W = copper sample plate holder (adapted from reference 8)

crons but slow spray on rates are generally needed to produce a sufficiently low scattering matrix. Depolarisation ratios can in principle be obtained. Different areas of the matrix can be studied and with molecular matrices like nitrogen, the matrix vibrations can also be monitored. However, fluorescence from the sample or from impurities can give the usual problems and rather high sample/matrix concentration ratios need to be used because of the inherent weakness of the Raman effect.

5.3 High Pressure

Application of high pressures to liquids and solids can produce structural changes, shifts in reaction and conformational equilibria and will modify molecular dynamic behaviour. In many ways Raman spectroscopy is the more effective way of studying vibrational spectra under these conditions. It requires window transparency in the visible region and thus a complete vibrational spectrum can be obtained using diamond, sapphire, quartz or glass windowed cells. Detailed reviews of the techniques involved have been published [11, 12].

There are two principal types of cell which are commonly used — anvil cells and hydrostatic equilibrium cells. Diamond anvil cells are used for very high pressures, these rely on the application of a moderate force using the mechanical advantage of a lever to squeeze a sample between the optically flat and aligned faces of two diamonds having a small surface area and thus generating a high pressure. Pressures generated in this way can be very high but result in an approximately parabolic pressure distribution across the sample fading to zero at the face edges. This pressure variation can be minimised by using a metal gasket (inconel is often employed) between the diamond faces to retain the sample. This improvement also allows liquid samples to be studied and for hydrostatic pressures to be applied to solids suspended in a fluid. In this way pressures up to 60 or 70 kbar using 1.2 mm diameter faces and 400 kbar using 0.4 mm diameter faces can be easily achieved. With even smaller faces and very careful alignment pressures up to 1.7 Mbar have been reported. The holes in the sample retaining gaskets have to be carefully made and positioned and should be less than half the diameter of the high pressure face.

A typical design is shown in Fig. 5.7 which uses a small hydraulic ram giving better control over the applied pressure to the two opposed diamonds than the alternative spring washer loading system. For lower pressures one of the anvils can be replaced by a tungsten carbide plate and 180° scattering can be used. Glass anvils have also been used for pressures up to 12.6 kbar [13].

Various illumination geometries have been used with diamond anvil cells. Good results can be obtained using 0° scattering with an elliptical mirror to collect the Raman light [14]. Alternatively 180° scattering can be used with conventional slit optics [15] or most conveniently by employing a long working distance objective with a Raman microscope [16]. This latter method has been used extensively by the author for studies on liquids and solids and simplifies focusing on the small sample volume, whilst rendering efficient collection of the scattered light.

Accurate pressure measurement in an anvil cell is difficult. The stress induced shift of the diamond Raman signal can be a useful guide but does not really reflect the

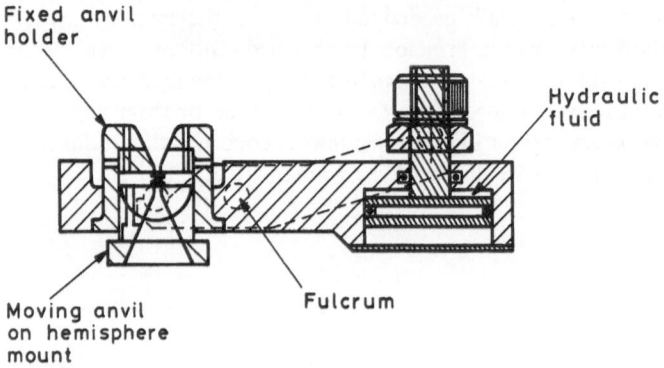

Fixed anvil
holder

Hydraulic
fluid

Moving anvil
on hemisphere
mount

Fulcrum

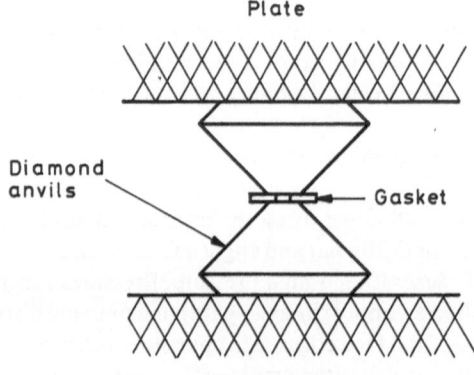

Plate

Diamond
anvils

Gasket

Hemisphere

Fig. 5.7. Compact, hydraulic, diamond anvil cell, produced by Diacell Products.

pressure experienced at the sample. Fluids whose pressure induced Raman shifts are known can be used as internal calibrants and the frequency shift of the anthracene absorbance around 376 nm has also been employed [17]. However the most widely used method relies upon the pressure induced shift of the ruby fluorescence bands. A very small fragment of ruby is added to the sample volume and the R1 band of the fluorescence spectrum at 694.2 nm which shifts by 0.76 cm^{-1} kbar^{-1} is monitored during the experiment [18].

Hydrostatic cells, as the name suggests are essentially pressure vessels fitted with windows and filled with a hydraulic fluid medium which is pressurised by an external pump. The hydraulic fluid may either be used to apply a hydrostatic pressure to a solid sample located within the cell or may be the subject of the experiment itself. Alternatively, gas pressure applied via a pressure intensifier can be used and, if necessary, the hydraulic fluid can be separated from the test fluid by means of a phase separator. Most commonly, a 90° scattering geometry is used which employs three windows. Two windows in line allow the laser beam to be focused into the cell and to be refocused on the exit side back into the cell to give a double pass. The third window is set at 90° to the laser illumination direction and opposite the mid point between the first two mirrors to allow collection of the scattered light. The design of the window housing is critical in order to achieve a high pressure seal. The apertures of the illu-

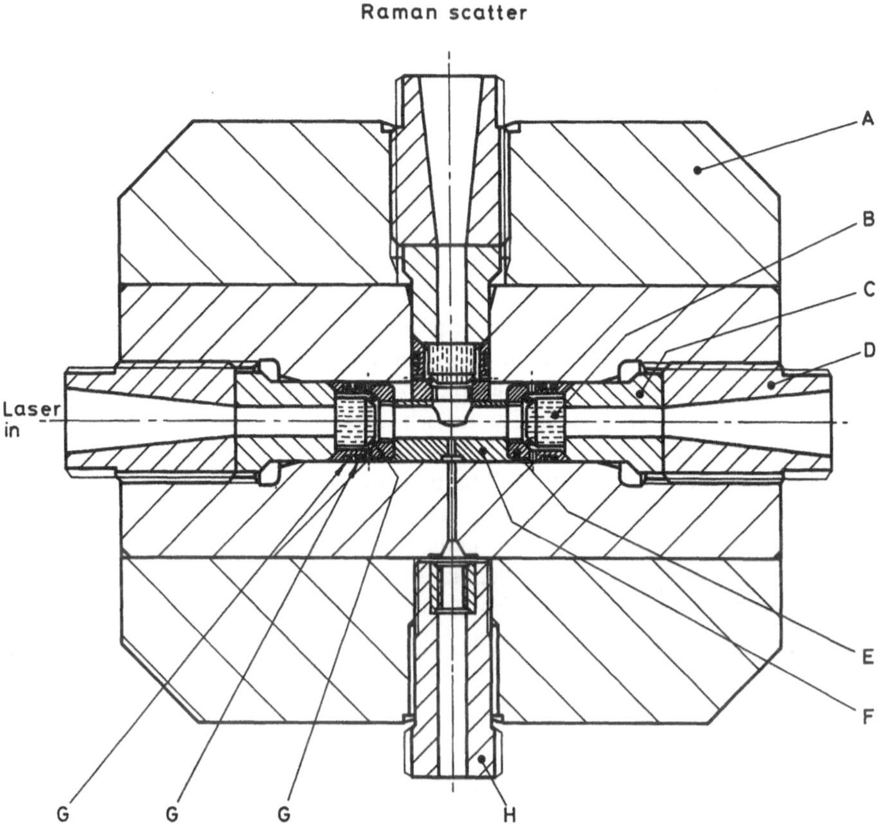

Raman scatter

Fig. 5.8. High pressure Raman cell produced by Nova Swiss Ltd. Contstructed from 316 stainless steel and capable of operating at pressures up to 7 kbar. A = stainless steel body, B = sapphire window, C = window support, D = screw-in window plug, E = copper window holder, F = inner cell, G = O-rings, H = sample fluid and hydraulic inlet

mination windows can be small as they need only accept a focused laser beam, but the aperture and indeed the exit cone defined by the window housing of the scattered light window, needs to be maximised within the pressure constraints of the cell, in order to observe maximum scattered light intensity. Figure 5.8 shows a typical hydrostatic high pressure cell available commercially from Nova Swiss Ltd. which can operate at up to 10 kbar and is pressurised by a simple piston-in-cylinder hand pump.

A solid medium can also be used in high pressure cells. In this case the cell is filled with the pressure transmitting solid, usually sodium chloride, and a piston forced into the cell generates the pressure. This is the basis of the Drickamer type high pressure cell.

Pressure calibration below 10 kbar is somewhat easier in the hydrostatic cells as the sample volume is very much larger than in the case of the diamond anvil cells. Bourdon gauges, manganin resistance gauges and precalibrated strain gauges fitted to the cells or to the inlet high pressure piping have all been used. At higher pressures it is more convenient to use internal standards having known pressure shifts or to rely on the ruby fluorescence method.

In many experiments it is important to be able to vary the temperature in addition to the pressure in order to investigate temperature effects at constant density. External electrical heating can be used to achieve temperatures up to around 500 °C which can be controlled and monitored as described in Sect. 5.1. Low temperatures can also be attained by passing coolant through the cell body or for very low temperatures hydrostatic cells can be attached to a liquid helium cryostat [19]. Diamond anvil cells can be used whilst immersed in liquid nitrogen or helium cryostats [20].

Many aspects of Raman spectroscopy require a careful analysis of the polarisation characteristics of the scattered light. Stress induced birefringence in the window materials of high pressure cells will alter the polarisation properties of both the incident light and the scattered light. Careful selection of window material can be of assistance. For example the use of float glass, which has been used successfully up to 45 kbar [21], will minimise the effect. Alternatively, the polarisation analysis can be achieved by placing the polaroid filter inside the cell, any subsequent scrambling is then unimportant. The effects of stress induced birefringence can be quantified and hence allowed for by measuring the depolarisation ratio for the Raman band of an approximately spherical molecule which can be assumed to be little affected by pressure [22]. The scrambling of polarisation by glass windows increases quadratically up to 18 kbar, but this increases markedly away from the central axis of the window. Nevertheless, for glass windows, it has been calculated that the degree of scrambling is $<10^{-3}$ at 18 kbar [23].

5.4 Special Chemical Environments

5.4.1 Corrosive Samples

Raman spectroscopy is particularly suited to the study of the vibrational spectra of corrosive liquids. That is to say liquids which would react with or dissolve normal infrared cell windows. For many purposes Pyrex glass or quartz is sufficiently unreactive to retain many reactive samples like concentrated acids and oxidising agents. Liquids like HF solution, CCl_3, BrF_3, BrF_5, O_2F_2 and concentrated H_2SO_4 and HNO_3 have all been studied in this way. Corrosive melts as mentioned in Sect. 5.1 can also be accommodated using windowless cells.

5.4.2 Thin Films and Interfaces

Studies of gas-solid, liquid-solid, liquid-liquid and solid-solid interfaces are of great importance both theoretically and industrially. For example in heterogeneous catalysis, gas phase metal corrosion and reactions at electrode surfaces the nature of the interaction between the film or adsorbed species and the substrate is critical both from a structural and an energetic standpoint. One important practical approach to this type of study is the use of integrated optics or wave guiding techniques which essentially establish a multipassing illumination geometry through multiple internal reflections in the film in order to enhance the Raman intensity. This approach is discussed in detail in Chap. 2.

5.4.2.1 Gas-Solid Interfaces

The major problems associated with the study of adsorbed gases by Raman spectroscopy is the low concentration of scatterers and the relatively high intensity of the specularly reflected exciting light and any associated fluorescence. Various methods have been tried to eliminate the problems for specific applications. The concentration problem can be accommodated using a powdered adsorbate thus increasing the surface area available for adsorption. However, in the case of zeolites this approach is generally associated with fluorescence problems which can be tackled using temporal fluorescence rejection techniques [24]. In suitable circumstances it may also be possible to use wave guiding techniques to achieve a multipass of the exciting laser through the adsorbed film. When a plane adsorbate surface is being studied, it is possible to arrange the scattering geometry to optimise the Raman scatter-to-specular reflection ratio. Rejection of the specular reflection reduces the random scatter in the spectrometer resulting in low background spectra in which very low photon count signals can then be observed. Unenhanced Raman spectra have been observed in this way under ultra-high vacuum from 0.5 monolayer coverages of pyridine on nickel surfaces [25].

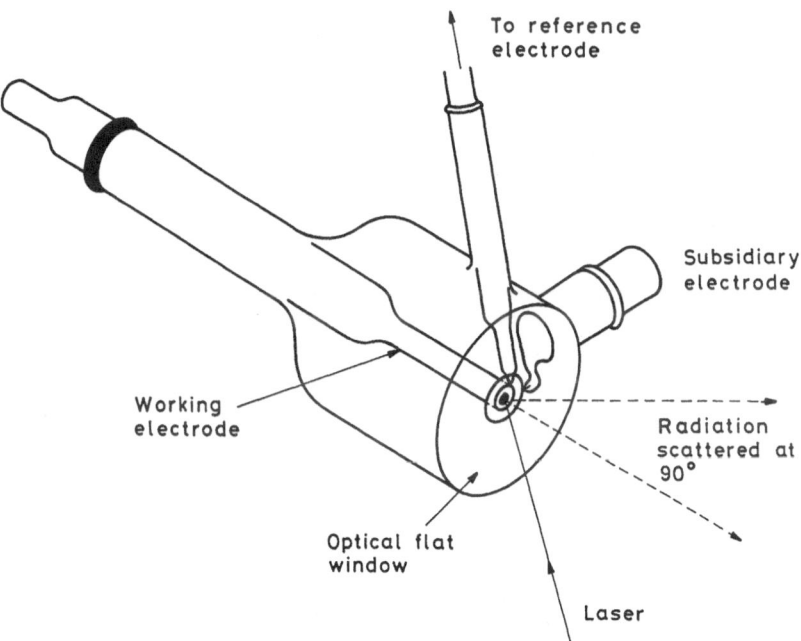

Fig. 5.9. Essential details of an ambient temperature and pressure electrochemical cell (Reproduced from reference 24 with permission)

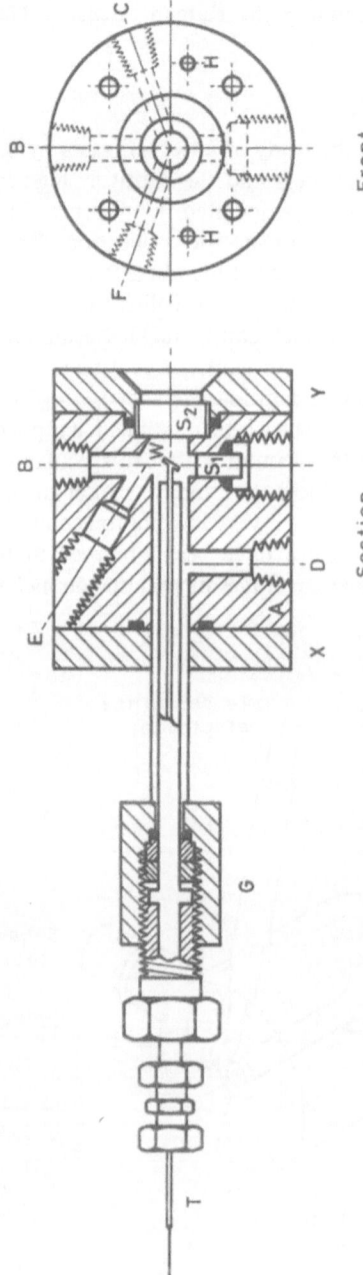

Fig. 5.10. Cell for studying electrode surfaces at high temperatures and high pressures. See text for explanation. (reproduced from reference 24 with permission)

5.4.2.2 Liquid-Solid Interfaces

Electrochemical cells provide a particular interest area in this type of interface. It is fairly straightfoward to construct a cell to study electrode surfaces at ambient conditions as all that is required is a sealed glass cell in which the working electrode can be positioned close to a flat region of the cell to allow the laser beam to be focused on the surface of the electrode and to allow the scattered Raman light to be collected. A design [26] of such a cell is shown in Fig. 5.9. The working electrode is made from 4.5 mm diameter silver rod, sheathed in Kel-F, which can be repolished after each experiment. The subsidiary electrode, made from a 1 cm diameter platinum wire ring and the Luggin capilliary of the reference calomel electrode, are positioned just outside the scattering area. Best spectra were obtained with the laser incident at an angle of 68° to the axis of the working electrode and with the electrode surface 5 mm behind the optical flat. Many cells of this type have been used to study the surface enhanced Raman scattering effect in which some species adsorbed at specially pre-pared silver electrodes give rise to greatly enhanced Raman intensities. The prospect of examining an electrochemical corrosion process in situ at high temperatures and pressures, presents additional experimental problems. Figure 5.10 shows the cell used by Melendres et al. [27] to study the corrosion and passivation of lead in very dilute sulphate solution from 25 °C to 290 °C at 100 bar pressure. The design contains all the essential features for this type of work and could be adapted to suit most instrument arrangements. The electrolyte solution is preheated and continuously pumped through the cell, entering at port D and leaving at port C. The counter electrode is inserted through E and the cell is heated by heating cartridges H, embedded in the cell body. A shaft seal G allows positioning of the electrode and the electrode leads are brought out at T.

5.4.2.3 Liquid-Liquid Interfaces

Mechanisms of surfactant behaviour and of micelle formation are amongst the interest areas involving this type of interface. The first observation of the Raman spectrum from a monolayer interface between two liquids was recorded by Takenaka and Nakanaga [28]. This was obtained from a monolayer of the product formed from cetyltrimethylammonium bromide (a cationic surfactant) and methyl orange in aqueous solution at an interface with carbontetrachloride. The product gave rise to a resonance Raman spectrum when excited by total reflection of a laser beam at the interface.

5.4.2.4 Solid-Solid Interfaces

There are extensive areas of research involving this type of interface including multi-plex thin films, semiconductor devices and ceramics. One interesting example which can be regarded as belonging to this category is the Langmuir-Blodgett film which can be produced precisely at mono and multi-layer thicknesses. Recent work on such films [29] has taken advantage of the excellent low noise characteristics of a specially prepared charge coupled device detector, described in Chapter 3, to observe the spectra of single and multiple layers of cadmium stearate deposited on a glass slide. The spectral quality is excellent and use of this type of detector may well be the way forward in extending the study of interfaces by Raman spectroscopy.

5.4.3 Light Absorbing and Thermally Sensitive Samples

Raman Spectroscopy can present difficulties when studying photochemically and thermally sensitive samples. This is necessarily the situation in the case of resonance Raman spectroscopy, where intensity enhancement due to a resonance mechanism requires that the laser wavelength must lie within a strong electronic absorption profile of the sample (see Chapter 1). The consequences are risk of thermal degradation of the sample and attenuation of the scattered light through re-absorption by the sample. Sample cooling (see earlier) can be used to reduce the thermal response of this type of sample but in most situations the most effective approach is to use a cell which spins the sample thereby reducing the residence time of the laser on any one spot of the sample. Amongst the first cells to be employed using this technique are

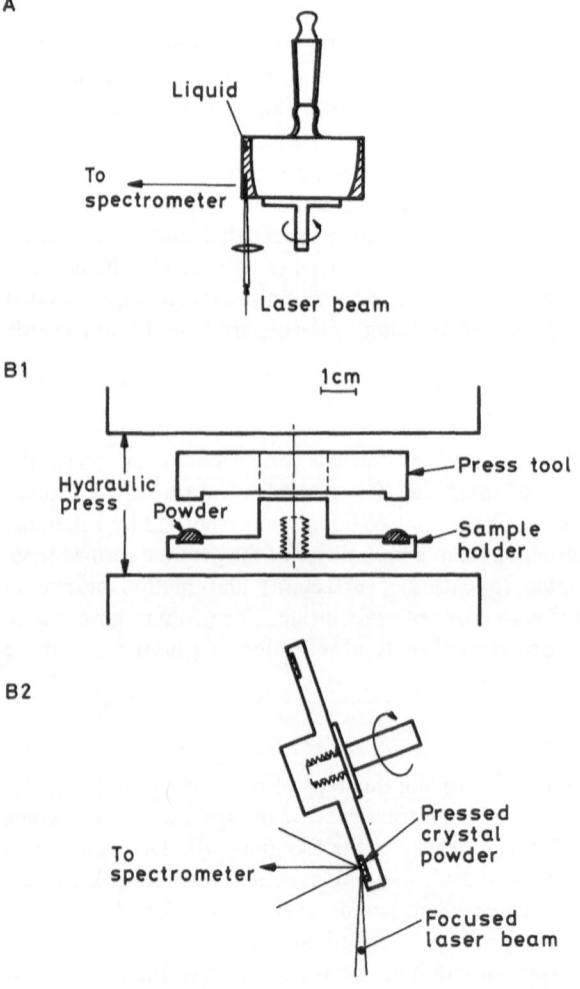

Fig. 5.11. Rotating Raman cells. A for use with liquids, B for use with solids. B1 shows the method for loading the cell, B2 shows the alignment geometry. (Reproduced from reference 30 with permission)

those described by Bernstien and Kiefer [30] which are shown in Fig. 5.11. The liquid cell is made from quartz or Pyrex glass which is glued to a brass rod which in turn fits into the chuck of an electric motor rotating at speeds around 1000 rpm. It is important to position the laser beam close to the side of the cell to minimise self-absorption of the laser scattered light. The solid cell is loaded by filling the annular ring with powdered sample and then using the tool to press the sample into a compact form which will remain fixed in the cell during rotation. An alternative method which has been used successfully by the author, is to press the sample into a disk using an infrared sample press, using a small amount of KBr to assist cohesion, when necessary. The resulting sample disk can then be spun in the laser beam by attaching it with double sided adhesive tape to the spindle of a small electric motor. Several variations of these cells have been used including cells which can operate in a back scattering geometry [31], at low temperatures [32] and under vacuum [33].

5.5 References

1. Gardiner DJ, Jackson RW, Straughan BP (1979) J. Mol. Struct. 54: 31
2. Gardiner DJ, Littleton CJ, Walker NA (1987) J. Raman Spectrosc. 18: 9
3. Harley RW, Manning DI, Ryan JF (1978) J. Phys. E 11: 517
4. Gilbert B, Mamantov G (1975) J. Chem. Phys. 62: 950
5. Kawamura K, Okada I (1976) Rev. Sci. Instrum. 47: 384
6. Gardiner DJ, Turner JJ (1971) J. Mol. Spectrosc. 38: 428
7. Gans P, Gill JB (1976) J. Phys. E 9: 301
8. Downs AJ, Hawkins M (1983) Adv. Infrared and Raman Spectrosc. 10: 1
9. Jodl HJ (1984) In: Durig JR (ed) Vib. spectra and structure, Elsevier, Netherlands, vol 13, chap 6
10 Shirk JS, Classen HH (1971) J. Chem. Phys. 54: 3237
11. Sherman WF, Wilkinson GR (1980) In: Clark RJH, Hester RE (eds) Advances in infrared and Raman spectroscopy, Heyden, UK, vol 6, chap 4
12. Gardiner DJ (1982) Roy. Soc. Chem. Ann. Rep. C Ch 5
13. Wong PTT, Moffatt DJ (1983) Appl. Spectrosc. 37: 85
14. Sharma SK (1978) Year Book, Carnegie Institution: 660
15. Schoen PE, Schnur JM, Sheridan P (1977) Appl. Spectrosc. 31: 337
16. Gardiner DJ, Bowden M, Daymond J, Dare-Edwards MP (1984) Appl. Spectrosc. 38: 313
17. Hill RM, Christian SD (1982) Appl. Spectrosc. 36: 302
18. Piermarini GJ, Block S, Barnett JD, Forman RA (1975) J. Appl. Phys. 46: 2774
19. Jodl HJ, Holzapfel WB (1979) Rev. Sci. Instrum. 53: 341
20. Silvera IF (1985) Rev. Sci. Instrum. 56: 121
21. Whalley E, Lavergne A, Wong PTT (1976) Rev. Sci. Instrum. 47: 845
22. Perry S, Sharko PT, Jonas J (1983) J. Appl. Spectrosc. 37: 340
23. Ikawa S, Whalley E (1984) Rev. Sci. Instrum. 55: 1273
24. Howard J, Everall NJ, Jackson RW, Hutchinson K (1986) J. Phys. E: Sci. Instrum. 19: 934
25. Harradine D, Campion A (1987) Chem. Phys. Lett. 135: 501
26. McQuillan AJ, Hendra PJ, Fleischmann M (1975) J. Electroanal. Chem. 65: 933
27. Melendres CA, McHahon, W. Ruther W (1986) J. Electroanal Chem. 208: 175
28. Takenaka T, Nakanaga T (1976) J. Phys Chem. 80: 475
29. Murray CA, Dierker SB, (1986) J. Opt. Soc. Am. A 3: 2151
30. Kiefer W, Bernstein HJ (1971) Appl. Spectrosc. 25: 500, 609
31. Rodgers EG, Strommen DP (1981) Appl. Spectrosc. 35: 215
32. Walters MA (1983) Appl. Spectrosc. 37: 299
33. Brown FR, Makovsky LE, Rhee KH, (1977) Appl. Spectrosc. 31: 563

CHAPTER 6

Raman Microscopy

by J. D. Louden

ICI Chemicals and Polymers Group, Characterisation and Measurement Group, Research and Technology Department, The Heath, Runcorn, Cheshire, WA7 4QD, England

6.1 Introduction

In 1966, Delhaye and Migeon suggested that laser-excited Raman scattering could be used in the analysis of microscopic particles [1] and in an abstract published in 1973 [2]. Hirschfeld outlined some basic requirements for Raman microprobe spectrometers. The first results of practical Raman microscopy were reported at the IVth International Conference on Raman Spectroscopy in 1974 [3, 4] by two independent groups.

One group was from the National Bureau of Standards (NBS) in Washington (Rosasco et al.). The NBS system [5] consisted of a conventional laser Raman spectrometer in which the optics comprised a microscope objective to focus the laser beam down to approximately 1 μm, standard Raman collection optics to focus the magnified image onto the monochromator entrance slit and a separate 180° opposed microscope objective with which to view the sample in whitelight. This system was effective in demonstrating the feasibility of obtaining Raman spectra from micrometer and sub-micrometer particles but it had limitations in terms of ease of use, particularly sample mounting and alignment. The other group from the Laboratoire de Spectrochimie Infrarouge et Raman, Université de Lille, France (Delhaye and Dhamelincourt) described a micro-Raman spectrometer based around the tool of the microscopist, the optical microscope, to create a totally new instrument. The basic configuration of this instrument consists of coupling a conventional light microscope

Practical Raman Spectroscopy, Gardiner and Graves (Eds.)
© Springer-Verlag Berlin Heidelberg 1989

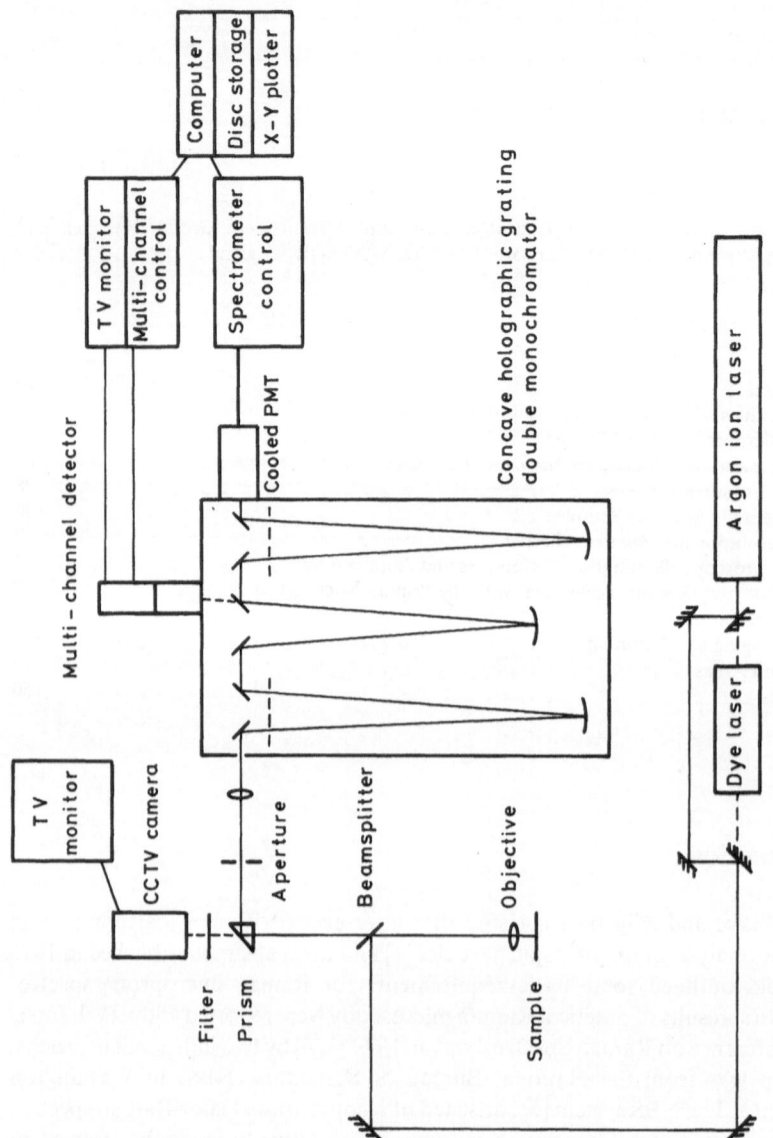

Fig. 6.1. Schematic of a microprobe system

to a double monochromator spectrometer equipped with concave holographic gratings, followed by two different detection systems (monochannel and multichannel). This system allowed single point analysis and Raman imaging/mapping. (Raman imaging will be discussed in more detail later in this chapter). This instrument was sold commercially as the MOLE (Molecular Optical Laser Examiner).

It is the single-point analysis Raman microprobes that have been used extensively in a wide range of research fields which will be described in this chapter. The construction of a microscope attachment for a Raman spectrometer was reported by Cook and Louden [6] and subsequently several manufacturers of Raman instrumentation offered microscope systems for attachment to any conventional Raman system.

6.2 Basic Principles of Microprobe Operation

A typical Raman microscope system is shown in Fig. 6.1. The sample to be examined is placed on the microscope sample stage and viewed using either transmitted white light for a transparent sample or incident white light for an opaque sample, on the viewing screen or closed circuit television viewing system. The area of interest of the sample is located centrally in the field of view and the sample irradiated by the laser beam.

The TEMoo visible laser beam from either an Argon or Krypton ion laser or dye laser is directed into the epi-illuminator of the optical microscope onto a semi-reflecting mirror (beamsplitter). A portion of the incident laser radiation is reflected downwards by the beamsplitter and a portion is transmitted through the beamsplitter. (The reflection/transmission characteristics of the beamsplitter is determined by the dielectric coatings used). The reflected beam then passes through the microscope objective lens which serves to focus the laser beam to a diffraction limited spot on the sample and to collect the scattered radiation from the sample. This scattered radiation is then transmitted through the beamsplitter to be viewed on the viewing screen or directed by a right angle prism and coupling optics to the entrance slit of the monochromator. The Raman signal is then detected either by the photomultiplier tube or the multichannel detector and the Raman spectrum then plotted on a chart recorder.

6.3 Raman Microprobe Optical Design

6.3.1 Microscope Choice

Raman microscopes are constructed from research grade metallurgical optical microscopes, (with transmitted illumination facility), and designs based on infinity corrected systems (Olympus, Nachet, Leitz) or focused systems (Nikon, Zeiss) have been constructed.

Figure 6.2 illustrates two Raman microscope designs based on the two microscope systems.

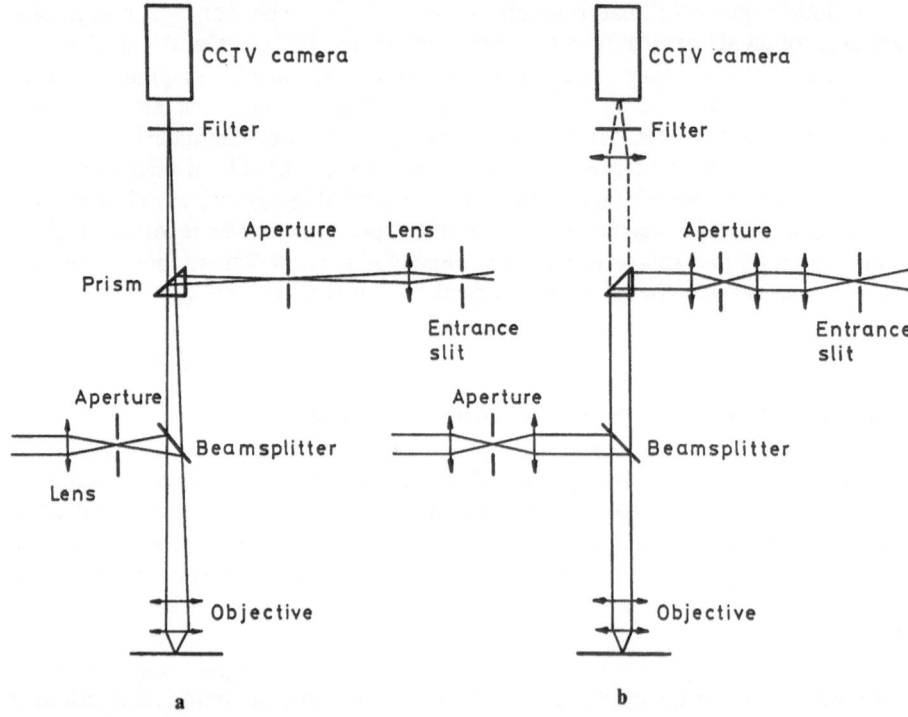

Fig. 6.2. Schematic of **a)** image plane microprobe system. **b)** infinity corrected microprobe system

When constructing a Raman microscope it is advantageous to use a microscope that is commonly used in the microscope department of your particular establishment. This enables various parts such as differential interference contrast equipment (DIC) and different objective lens etc, to be "borrowed" for special applications thus avoiding unnecessary duplication and expensive items purchases.

6.3.2 Beamsplitter Choice

The optical microscope is equipped with a dielectric beamsplitter with a 50/50 reflection/transmission ratio of incident unpolarised light. This beamsplitter is quite adequate for the majority of samples examined in Raman microscopy, but because laser power is not usually a limiting factor, an increase in sensitivity can be obtained by replacing the standard beamsplitter with a 10/90 or 20/80 reflection/transmission ratio. The available laser power has to be strongly attenuated (typically 0.5 to 20 mW) to prevent sample decomposition of thermally unstable samples, thus a 10/90 R/T beamsplitter decreases the laser energy impinging on the sample and allows the maximum scattered radiation energy to pass to the monochromator.

6.3.3 Objective Lens Choice

The choice of objectives depends on the type of samples to be examined. If the samples consist of microparticles in the sub-micron size range then a high magnification, high numerical aperture objective lens would be the lens of choice (a high numerical aperture objective (NA) will collect light scattered over a large solid angle so more Raman signal is detectable). A typical objective would be 100X magnification 0.9 NA or possibly 150X magnification 0.95 NA. If sample size is 5 microns upwards then a 50X or 60X 0.8 NA objective is adequate. If samples are buried inside host material then long working distance objectives are required. These are also useful for "in-situ" experiments. Generally a set of objectives which would cover most applications would consist of 150X 0.95 NA or 100X 0.9 NA for small samples (<1 micron → ~5 microns), 60X 0.8 NA or 50X 0.8 NA (5 microns upwards), 50X 0.55 NA ultra long working distance (8 mm working distance) for samples embedded in host materials, 20X 0.45 NA objective for liquid samples in glass capillary tubes, and a 10X 0.30 NA or 5X 0.1 NA for initial viewing and location of areas of interest.

Several manufacturers objectives have antireflection coatings that lead to spectral artifacts using certain laser lines, e.g. Nikon 40X and 100X and Olympus 80X, 100X

(A)

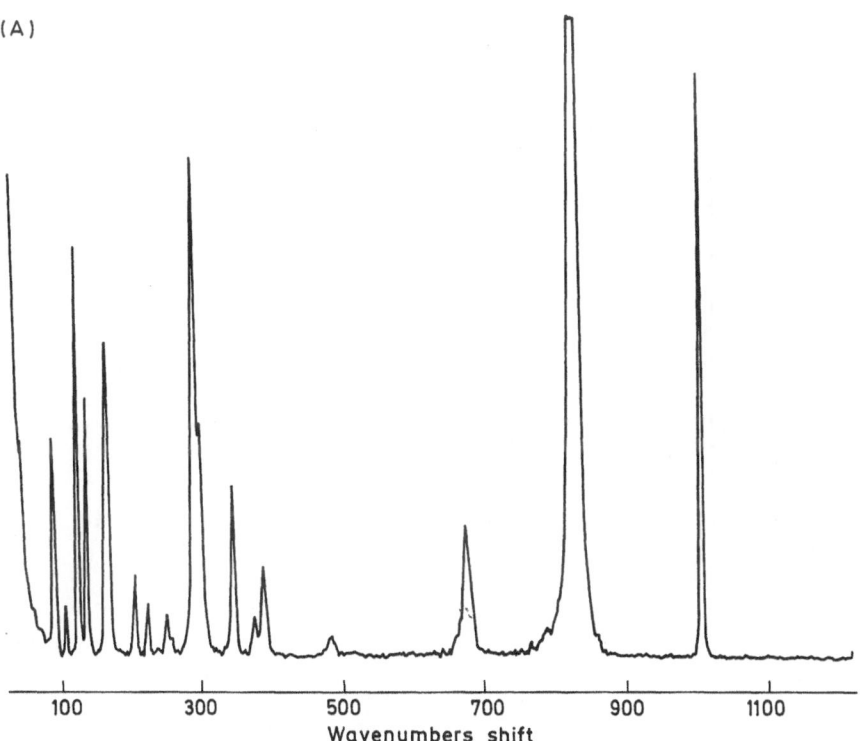

Wavenumbers shift

Fig. 6.3a. 10 μm × 10 μm crystal of MoO_3. 5 mW 488 nm at sample 60X 0.8 NA objektive. 1 sec TC 50 cm^{-1}/min scan. 50 μm slit width

and 150X give rise to peaks at 850 cm^{-1}, 910 cm^{-1}, 925 cm^{-1}, 1260 cm^{-1} and 1300 cm^{-1} when using the 514.5 mm argon ion laser line. These peaks do not appear using either the 488 nm or 647.1 nm laser lines indicating that they are a resonance Raman spectrum of the antireflection coating. However, they can be eliminated by using a spatial filter.

6.3.4 Viewing System Choice

The choice of viewing system is either a viewing (projection) screen or a closed circuit television system (CCTV). A viewing screen allows the operator to view the laser spot on the sample easily but when examining poorly reflecting samples in reflected white light illumination, the screen tends to be dark and it is difficult to distinguish features on the sample. Using a CCTV system on such a sample is much more convenient and the sample appears brighter and so features can be distinguished more easily. The drawback of the CCTV system is poor resolution of the vidicon camera tube (colour system) (High resolution black and white systems are available), and the

(B)

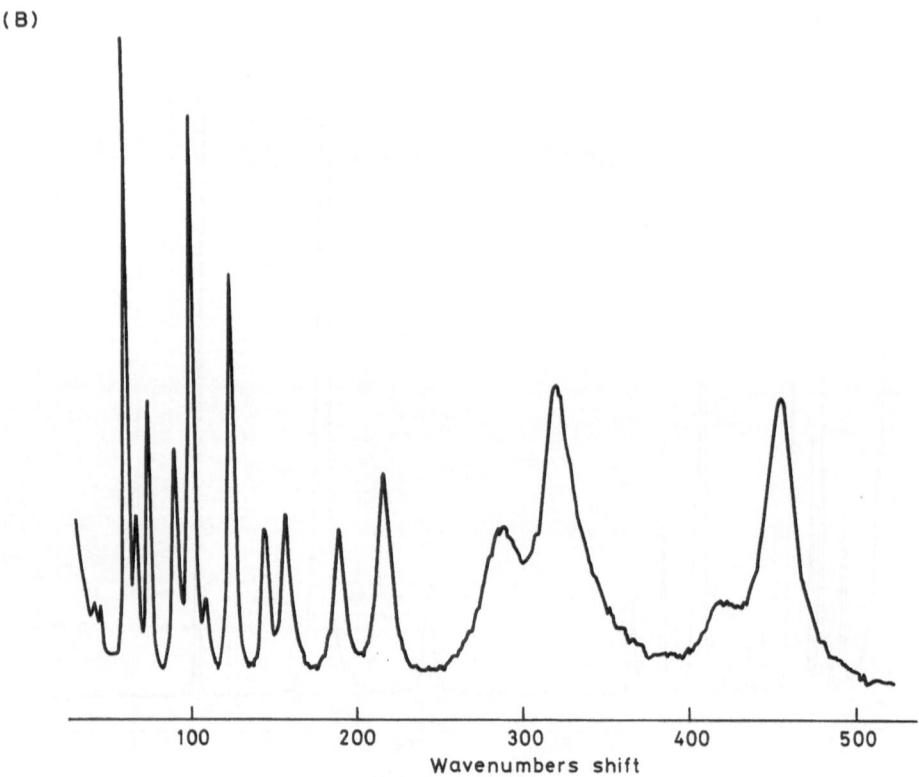

Wavenumbers shift

Fig. 6.3b. 10 × 10 μm crystal of Bi$_2$O$_3$. 5 mW 488 nm at sample 60X 0.8 NA objective. 1 sec TC 50 cm^{-1}/min scan. 100 μm slit width

laser spot cannot be viewed directly. A neutral density filter has to be incorporated before the camera tube to avoid damaging the tube. For general use a medium resolution colour camera closed circuit television system would be the best choice.

6.3.5 Coupling Optics

The coupling optics should consist of an achromatic doublet lens whose focal length was calculated so as to illuminate the whole width of the grating. The focal length of the coupling lens is much shorter than the tube length of the microscope, consequently the magnification factor of the microscope is reduced. In an achromatic doublet lens the chromatic aberration is cancelled at two distinct and well separated wavelengths (typically in the red and blue portions of the spectrum). Therefore the reflected laser light and shifted Raman scattered light are both focussed on the monochromator entrance slit. The calculation of a three-lens coupling optic is given by Dhamelincourt [7].

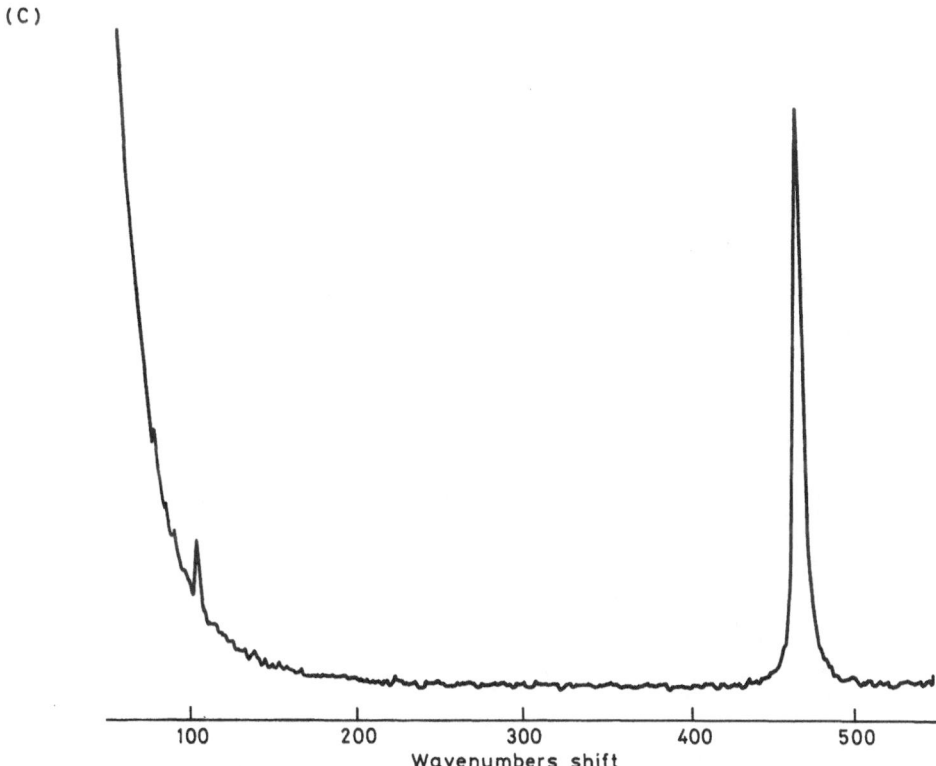

Fig. 6.3c. ~1 μm ThO$_2$ sphere. 5 mW 488 nm at sample 100X 0.9 NA objective. 1 sec TC 50 cm^{-1}/min scan. 250 μm slit width

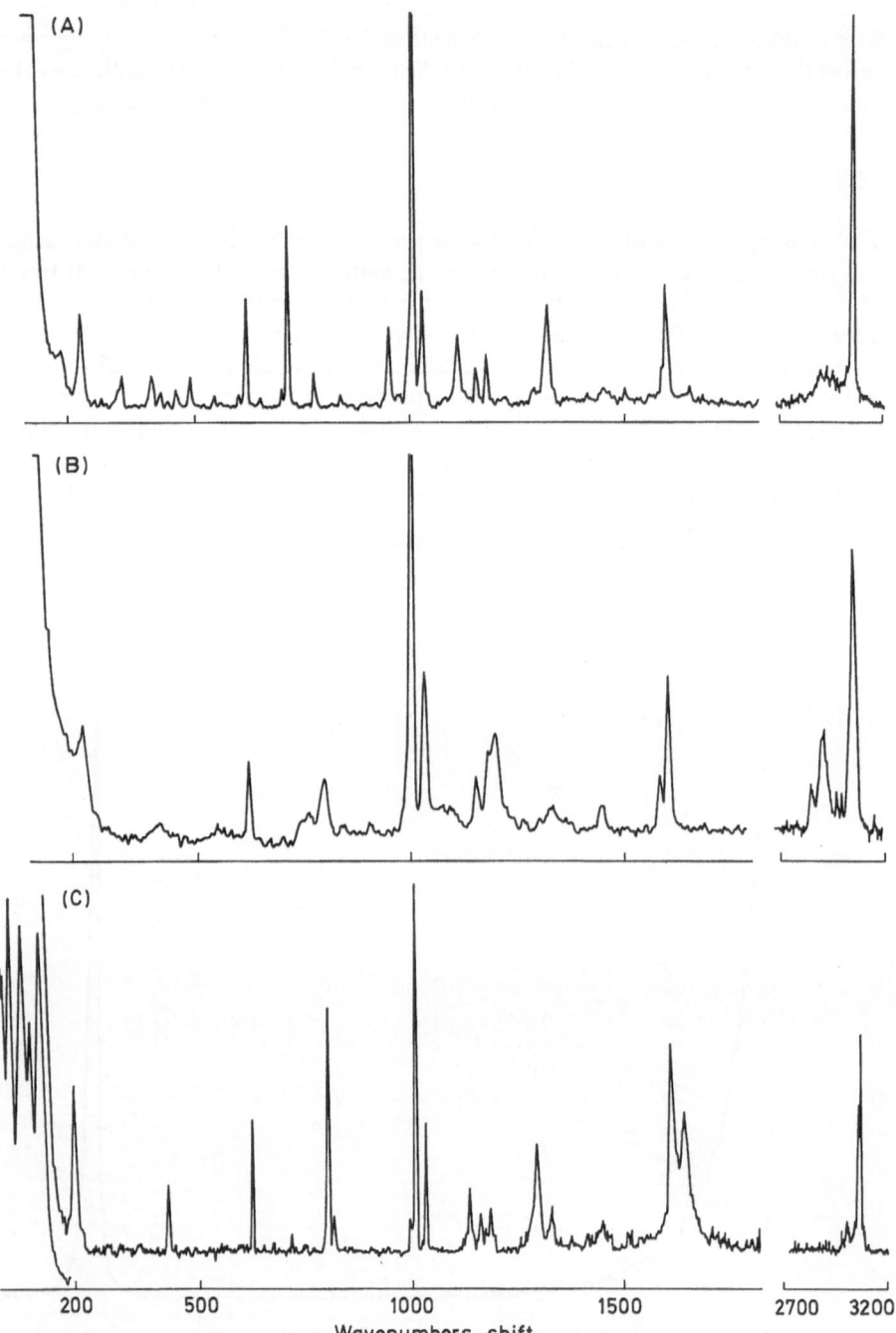

Wavenumbers shift

Fig. 6.4. A 15×10 μm Methyl centralite (NN^1 dimethyl, NN^1 diphenyl urea).
2 mW 488 nm at sample 60X 0.8 NA objective.
1 sec TC 50 cm^{-1}/min scan. 250 μm slit width. **B** 15×10 μm Polystyrene. 5 mW 488 nm at sample
60X 0.8 NA objective. 1 sec TC 50 cm^{-1}/min scan. 350 μm slit width. **C** 10×10 μm Benzoic Acid
0.5 mW 488 nm at sample 60X 0.8 NA objective. 1 sec TC 50 cm^{-1}/min scan. 250 μm slit width

6.4 Sample Preparation (Discrete Particles and Liquids)

The most convenient choice of sample support is an ordinary glass microscope slide, being both inert and a weak Raman scatterer, causing little or no interference to the Raman spectrum. Sapphire (α-Al_2O_3), LiF [8] and periclase (MgO) [9] have also been recommended.

The sample is simply placed on the glass slide, a suitable particle chosen and the spectrum recorded. Spectra of micron sized inorganic and organic species are easily obtained on the Raman microprobe. Inorganic species can be irradiated with quite

Table 6.1. Raman intensity signal of 459 cm^{-1} CCl$_4$ band with different objective lens

	Objective	NA	Counts sec^{-1} (30 mw 488 at Sple. 500 μm slit width)
Nikon	5×	0.1	1,170,000
	10×	0.25	1,500,000
	20×	0.4	1,380,000
	40×	0,55	1,130,000
	60×	0.8	1,126,000
	100×	0.9	700,000

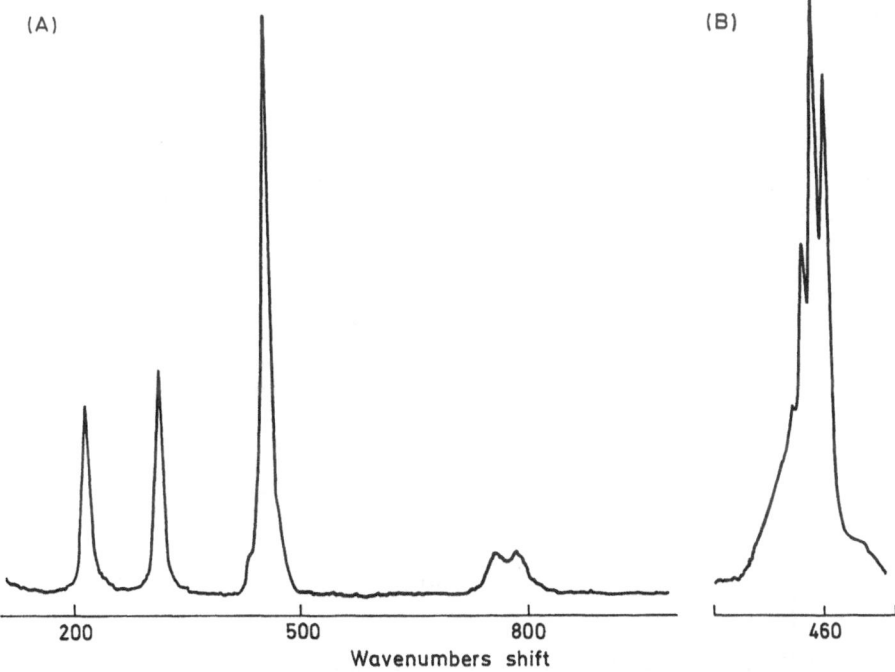

Fig. 6.5. Carbon Tetrachloride in capillary tube. A) 50 mW 488 nm at sample 20X 0.4 NA objective. 1 sec TC 50 cm^{-1}/min scan. 500 μm slit width. B) 459 cm^{-1} high resolution scan. 50 mW 488 nm at sample 20X 0.4 NA objective. 1 sec TC 5 cm^{-1}/min scan. 70 μm slit width

high laser power (typically 5 to 20 mW at the sample) without damage. Figure 6.3a–c shows spectra of inorganic materials of MoO_3, Bi_2O_3 and Tho_2 respectively. Figure 6.3c shows the capability of the Raman microprobe to obtain spectra of sub-micron particles. Organic species are more easily damaged and low laser powers are used to obtain spectra (typically 0.5 to 2 mW at sample). Figure 6.4a–c shows spectra of organic materials of methyl centralite (NN' dimethyl NN' diphenyl area), polystyrene bead and benzoic acid.

Recording of spectra of liquid samples is straightforward. The liquid is contained in a glass capillary tube (either flat or round thin-walled) and a low power objective (10X or 20X) used to record the spectrum.

Table 6.1 gives the Raman signal obtained using the 459 cm^{-1} carbon tetrachloride band with the different objectives. The CCl_4 was contained in a glass bottle one inch deep and covered with a coverslip. Figure 6.5 shows the Raman spectrum of carbon tetrachloride at low and high resolution.

6.5 Thermally Sensitive Samples

In Raman microscopy the laser beam is highly focused at the sample position to approximately 1 micron laterally, with a depth focus of between 3 and 10 microns (using high magnification, high NA objectives) producing high irradiance levels (typically 10^6 Wm^{-2}). Consequently, sample degradation can occur especially during long exposure times when recording the Raman spectrum.

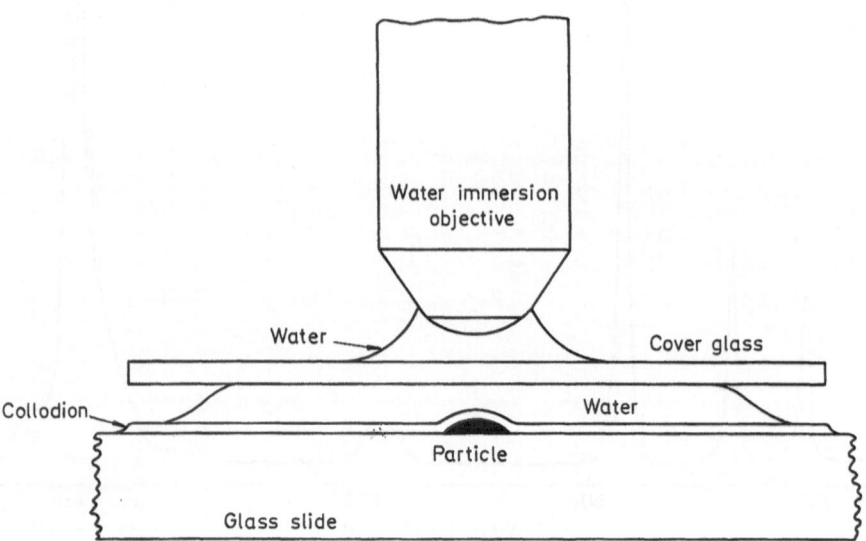

Fig. 6.6. Schematic showing the immersion technique as applied to small single particles (from Ref. 8). (Reprinted with permission from Analytical Chemistry (1981) 53: 1772. Copyright 1981 America Chemical Society)

One method to overcome this problem is to defocus the laser beam to reduce the power density especially when examining thermally sensitive compounds. Compounds which are effected by high-energy incident radiation (e.g. red compounds with a blue laser beam) can frequently be analysed by using lower frequency sources such as krypton ion or dye lasers.

Another technique is to immerse the sample in a suitable medium to act as a heat sink. Water is generally the ideal immersion liquid having a simple, weak spectrum and excellent thermal properties.

The particle is placed on the glass slide and immersed in a drop of water. Small particles can be held down using a thin layer of collodion to prevent their moving. Figure 6.6 shows the details. A water immersion objective can then be used to view the sample and record the spectrum.

Thermally sensitive but water soluble samples can be examined with the use of an organic immersion medium with which they do not react.

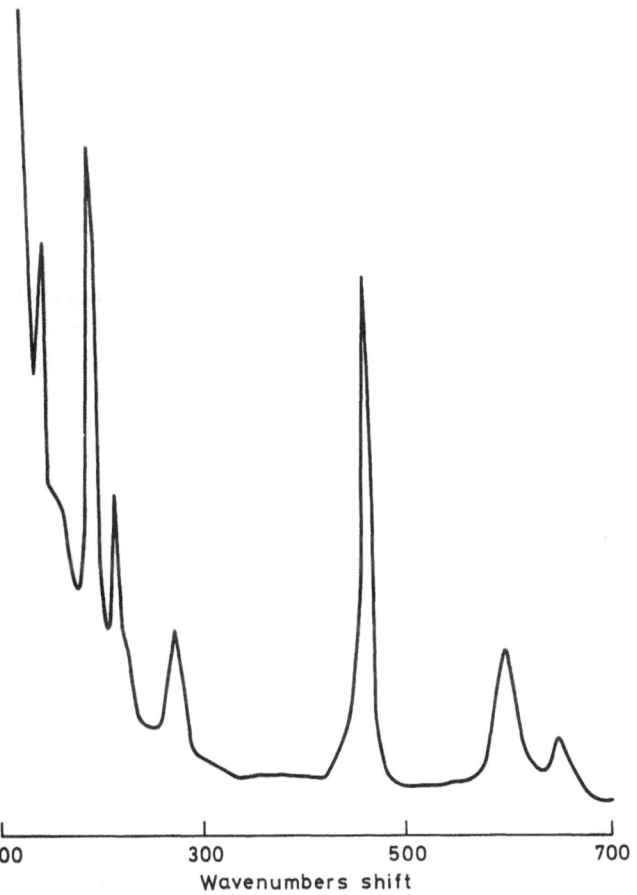

Fig. 6.7. Anhydrous CaCl$_2$ in sealed cell. (10 µm crystal). 5 mW 488 nm at sample 50X 0.55NA ultra-long working distance objective. 1 sec TC 100 cm^{-1}/min scan. 400 µm slit width

6.6 Environmentally Sensitive Samples

Environmentally sensitive samples can easily be examined by placing the sample inside a small flat-sided stoppered sample cell (similar to UV/Vis spectrometry cells) inside a dry-air glovebox, and the sample examined with an ultra long working distance objective through the walls of the cell. Figure 6.7 shows the spectrum of anhydrous calcium chloride contained in a glass cell filled with dry air. (Anhydrous calcium chloride is hygroscopic and rapidly dissolves in water absorbed from the air).

6.7 Spatial Resolution and Spatial Filtering

6.7.1 Axial Spatial Resolution

The incorporation of a spatial filter at an intermediate image plane between the microscope and monochromator serves to limit the depth of focus and field of view from which the Raman scattered light originates. With the use of the spatial filter, fluorescence, scattering, and Raman emission outside the laser focal volume are rejected by the intermediate aperture. Dhamelincourt [10] calculated that when the aperture size is equal to the lateral size of the image of the laser probe, the normalised depth (depth divided by refractive index) from which the Raman signal originates corresponds to about 2 to 3 μm above and below the plane of focus.

Adar and Clarke [11] presented an experimental arrangement to measure the axial spatial resolution of a microprobe system. The experimental arrangement is shown in Fig. 6.8. The sample consisted of a wedge of polycrystalline alumina fused to a lower wedge of polycrystalline Al_2O_3-6V/O ZrO_2 with the interface inclined at 6.5°.

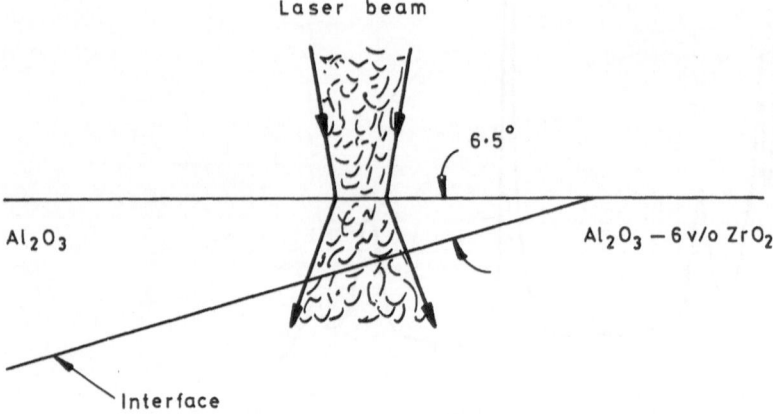

Laser beam

6·5°

Al_2O_3 $Al_2O_3 - 6$ v/o ZrO_2

Interface

Fig. 6.8. Schematic diagram of double-wedge poly crystalline sample used to evaluate axial spatial resolution of Raman microprobe (from Ref. 10).(Reprinted with permission from Microbeam Analysis (1982) p 308. Copyright 1982 San Fran cisco Press)

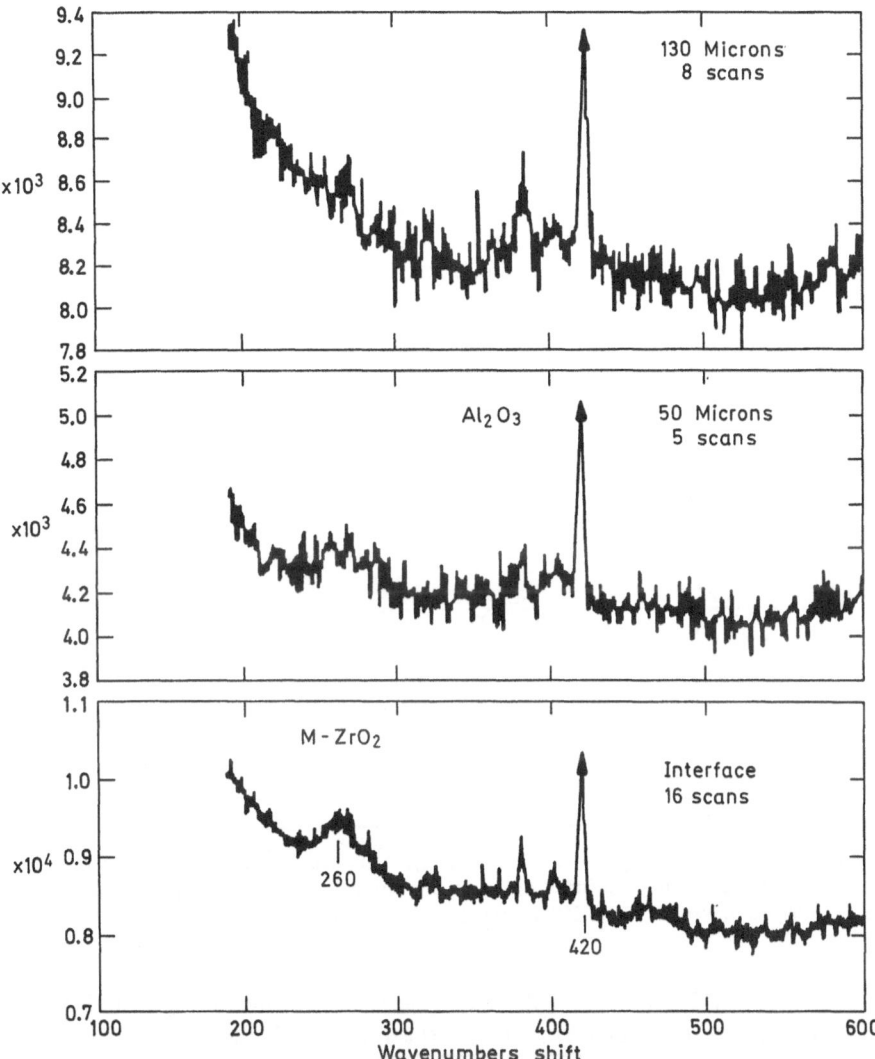

Fig. 6.9. Raman spectra recorded, with the geometry shown in 6.8, at Al_2O_3/Al_2O_3-6V/$OZrO_2$ interface, and at 50 and 130 µm laterally from interface; intensity of band at 260 cm^{-1}, characteristic of monoclinic ZrO_2, is a measure of depth resolution of microprobe. Wedge angle is 6.5° (from Ref. 10). (Reprinted with persission from Microbeam Analysis (1982) p. 308. Copyright 1982 San Francisco Press)

The laser was focused at successively greater distances from the interface and the Raman spectra obtained at each position. The intensity of the zirconia signal as a function of probe position may be used as a measure of the depth resolution of the microprobe. Spectra obtained are shown in Fig. 6.9. Using a 10 µm laser probe size and a 1 mm intermediate aperture a depth resolution of approximately 6 µm was determined. This result agreed with that predicted by Dhamelincourt.

6.7.2 Lateral Spatial Resolution

Adar and Clarke [11] also presented an experimental procedure for determining the lateral spatial resolution. The sample is shown in Fig. 6.10. Small ($\simeq 1\ \mu m$) grains of zirconia were embedded in a large alumina grain and the surface polished. Raman spectra were then observed at successive positions in the alumina grain up to the zirconia grain. No zirconia bands were detected until the laser probe was within half the probe diameter of the zirconia particle. Thus indicating a lateral spatial resolution of the order of 1 μm using a high numerical aperture (0.8 to 0.95 NA) objective lens.

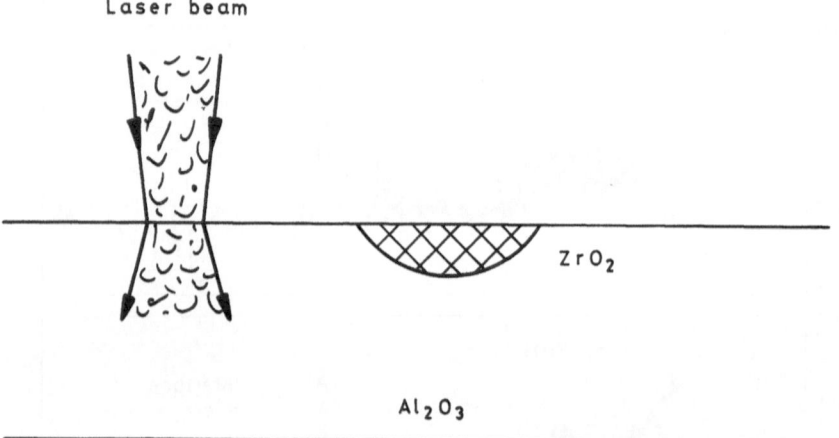

Laser beam

ZrO_2

Al_2O_3

Fig. 6.10. Schematic sample geometry for assessing lateral spatial geometry of polyphase, polycrystalline sample containing embedded phase particle (from Ref. 10). (Reprinted with permission from Microbeam Analysis (1982) p 309. Copyright 1982 San Francisco Press

Gardiner et al. [12] reported an alternative procedure for measuring spatial resolution using a highly reflecting surface with a sharply defined edge. The reflected laser light was measured at the coupling lens position simulating an idealized Raman experiment on an opaque sample, in which no spreading of the exciting radiation takes place. Edge responses were measured with and without a 100 micron spatial filter giving cut-off values of 0.9 microns and 0.6 microns respectively using a 100 × 0.9 NA objective lens. A simple expression for calculating beam waist is given in Sect. 2.4.

6.7.3 Spatial Filtering

The spatial filter allows the examination of samples where the region or particle to be studied cannot be physically isolated from the remainder of the sample, eg polymer laminates. Spectral interferences from the surrounding area can be minimised or eliminated by isolating the region of interest with the spatial filter. Figure 6.11 shows the Raman spectra obtained from a ∼50 μm laminate cross-section embedded in

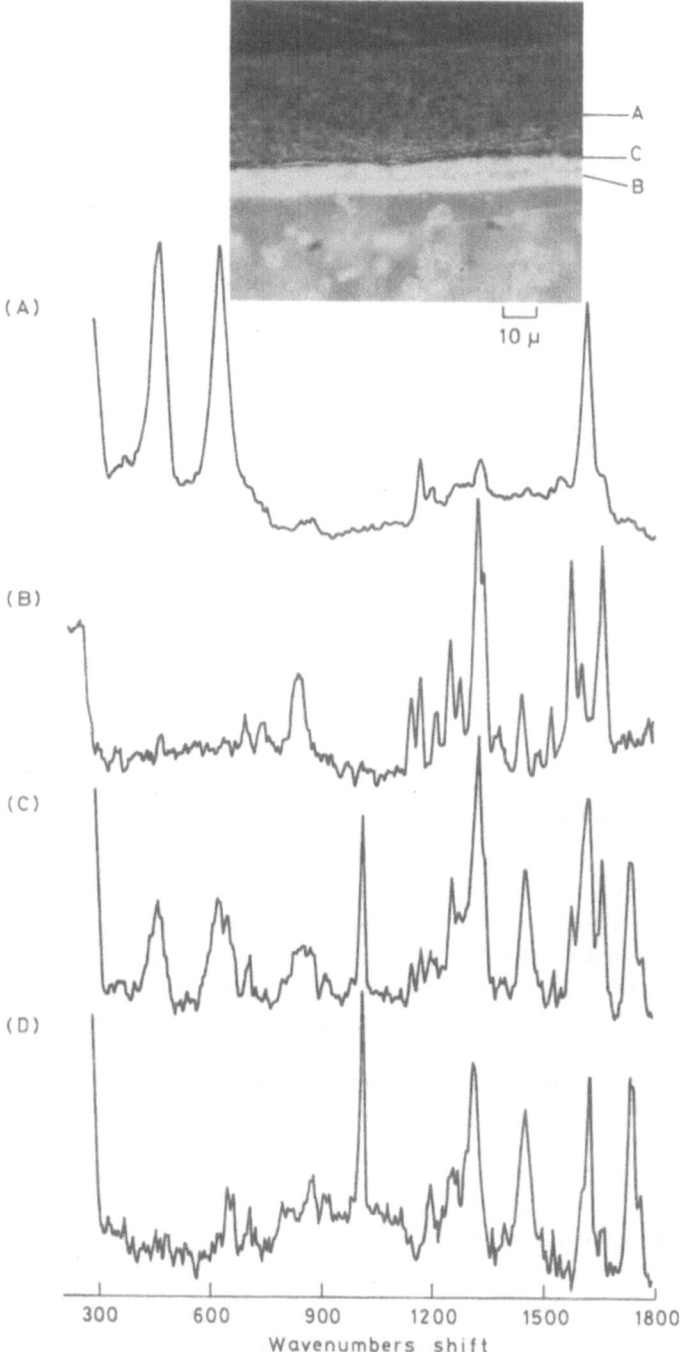

Fig. 6.11. Raman spectra of polymer laminate showing effect of spatial filter aperture. A) Outer 40 μm layer. B) Inner 10 μm layer. C) Middle 3 μm layer without spatial filter aperture. D) Middle 3 μm layer with 500 μm spatial filter aperture

epoxy resin. Spectra (a) and (b) show the outer 40 micron and 10 micron layers and spectra (c) and (d) were obtained from the inner 3 micron layer with and without the use of the spatial filter (aperture size 500 micron). The elimination of most spectral leakage from the nearby layers is indicated by the loss of Raman bands at 450, 650, 1575 and 1650 cm^{-1}.

6.8 Optical Microscope Illumination Techniques and Applications

The most common illumination systems in optical microscopes is transmitted bright-field and incident (reflected) bright-field. Transmitted bright-field is obviously used when examining transparent samples, such as polymer films, thin sections, biological specimens of cells and tissue etc, and incident bright-field is used to examine surfaces of solid objects such as corrosion on steel plates, rocks and stones etc. But the microscopist also has other illumination systems for looking at specimens enabling features to be detected that cannot be detected with bright-field illumination.

These other systems include polarised light (incident and transmitted), dark-field illumination (incident and transmitted), phase contrast (transmitted) and interference contrast illumination (incident/transmitted) and all these can be fitted to the Raman microprobe to enable visualisation and location of areas of interest to be examined by Raman microscopy.

6.8.1 Transmitted Polarised Light

Polarised light is used to detect strain patterns and birefringence in such materials as polymers, or crystals. A polariser is placed in the white light path before the sample and an analyser is placed after the sample. Figure 6.12 shows an irregularity in a polymer filament using white light and crossed-polarising filters. The distorted strain pattern and birefrigent streak are clearly distinguishable under these conditions but invisible under transmitted white light. The different areas can be probed with the laser and changes detected in the Raman spectrum examined. In this particular case the carbonyl (C=O) vibrational peak can be used as a measure of the crystallinity [13] of the different areas.

6.8.2 Incident (Reflected) Polarised Light

Similarly a polariser and analyser are placed in the white light path to distinguish strain in samples. Figure 6.13 shows a polymer film on a metal surface. Under incident white light the only features discernable are the lines on the metal surface. Under incident polarised light numerous additional features appear thus enabling spectra to be obtained from the different areas as shown in Fig. 6.13.

P.E.T. Filament bleb – white light

Cross polars

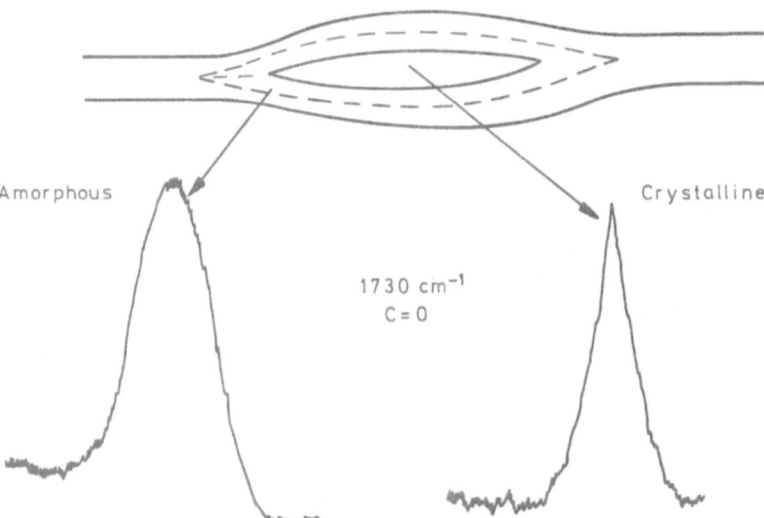

Fig. 6.12. Polyethylene Terephthalate Filament Bleb under transmitted white light and crossed polars. Strain pattern and birefringence only show under crossed polars

6.8.3 Incident (Reflected) Dark-Field Illumination

Incident dark-field illumination is carried out with dark-field objectives where the white light is sent down a separate path in the outer part of the objective (Fig. 6.14) to the specimen surface, producing polydimensional dark-field illumination. The optical elements of the objective do not contribute to specimen illumination. Only the wave fronts diffusely reflected by the specimen surface (reflectance) reach the

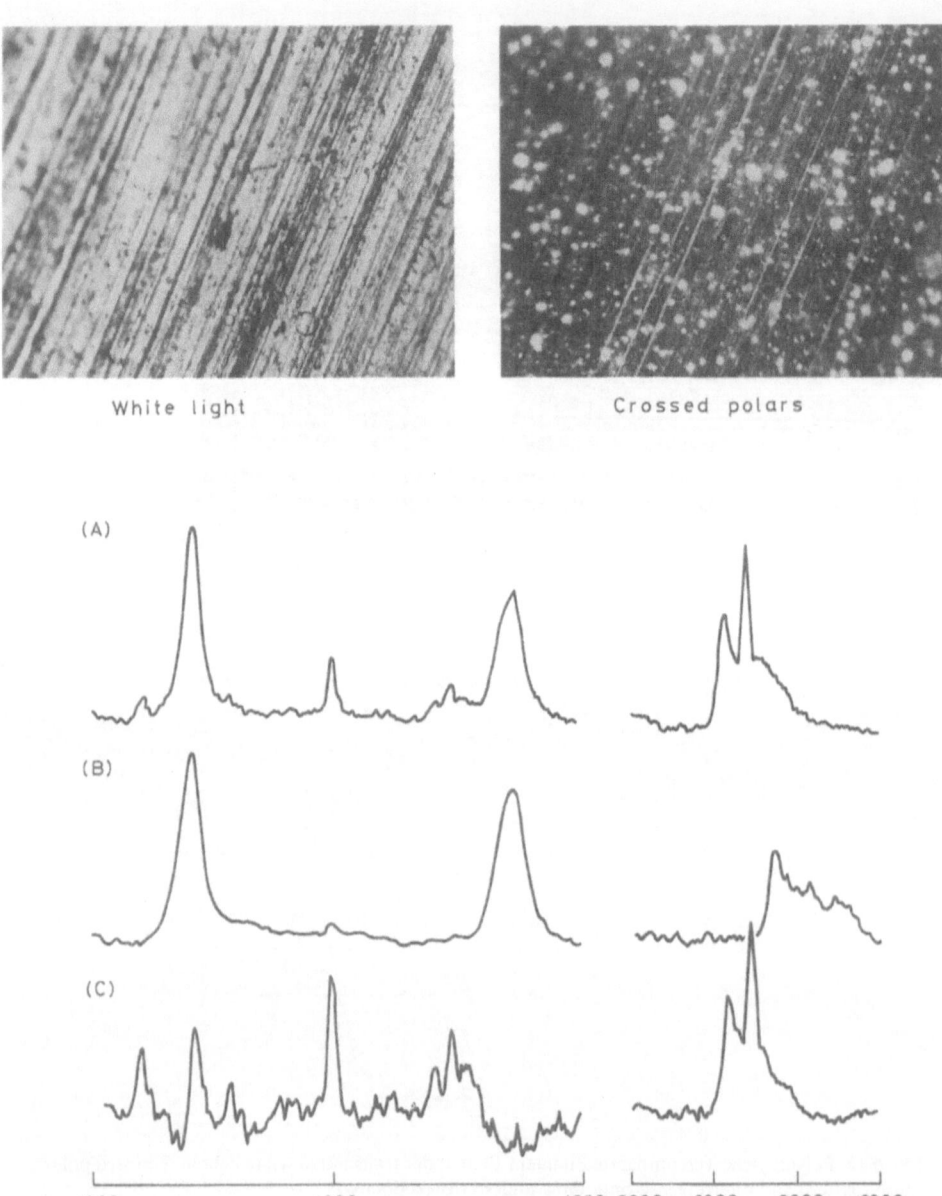

Fig. 6.13. Polymer film on metal surface seen under incident white light and incident crossed polars. Extra features are observed under crossed polars. A) white spots seen under crossed polars. B) Matrix area. C) Difference spectrum

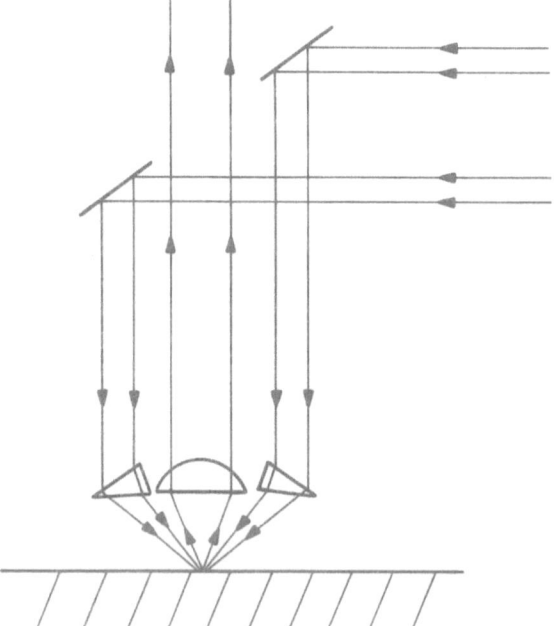

Fig. 6.14. Incident dark-field illumination path of optical microscope

objective and contribute to the dark-field image. Figure 6.15 shows photomicrographs of incident bright and dark-field illumination of a polymer laminate embedded in epoxy resin. In the bright-field image four different areas are distinguishable, but in the dark-field image a fifth very thin area (3 μm) is clearly seen in the middle. The associated Raman spectra are shown earlier in Fig. 6.11.

Incident bright - field   Incident dark - field

Fig. 6.15. Photomicrographs of polymer laminate under incident dark-field illumination. The middle (3 μm) layer is clearly visible under dark-field illumination

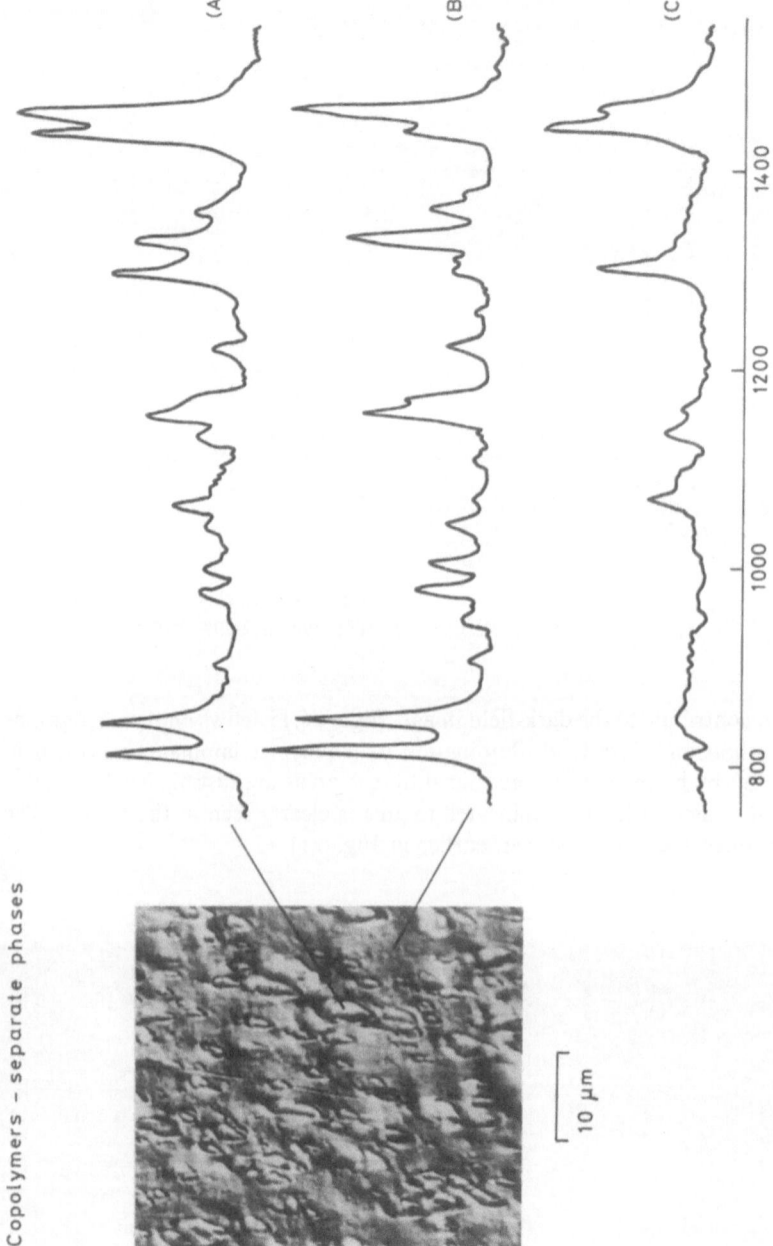

Fig. 6.16. Copolymer surface viewed under incident differential interference contrast technique showing separated phase structure. A) Separated phase. B) Matrix. C) Difference spectrum. (Reprinted with permission from Microbeam Analysis (1982) p 296. Copyright 1982 San Francisco Press)

6.8.4 Incident (Reflected) Differential Interference Contrast Illumination (DIC)

Under the ordinary metallurgical microscope a sample which has the same reflecting index throughout will not show minute unevenness or contour of the surface of the subject. However, if a polarisation type differential interference microscope is used, where the polarised light is applied and the lateral shift (shear) of two interfering light waves is limited within the resolving power of the objective, then minute surface structures will be made visible as an interference image of high contrast, thus allowing their detection. Figure 6.16 shows the photomicrograph of a polymer blend surface using DIC illumination. Different areas are clearly seen allowing Raman spectra to be obtained and so the differences in polymer composition can be determined.

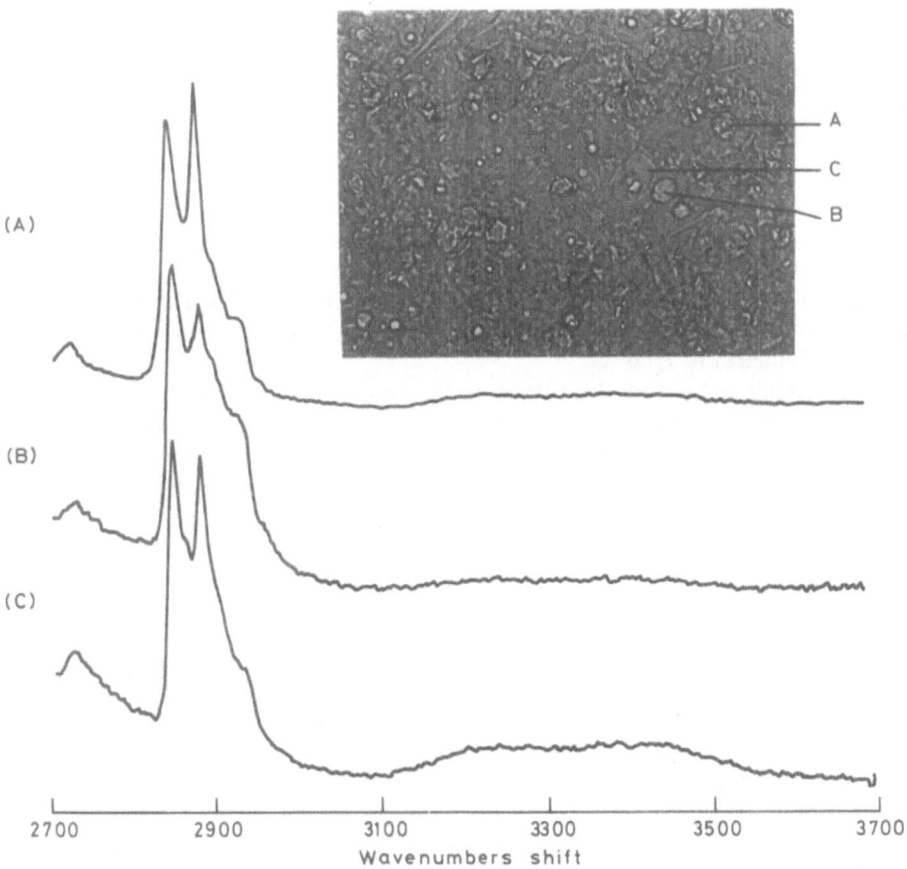

Fig. 6.17. Cetostearyl alcohol/cetrimide/water emulsion under transmitted differential interference contrast technique showing different phase structures and their associated Raman spectra

6.8.5 Transmitted Differential Interference Contrast (DIC) Illumination

In transmitted DIC the components of a specimen which have similar refractive indices are seen more clearly than with transmitted bright-field microscopy. The advantage of the method is that DIC essentially focuses on the ability to render surfaces of objects three-dimensionally. Therefore, the objects and their fine-structure details appear in relief-like contrast. With this technique object structures can be shown which cannot be recognised sufficiently clearly with the conventional methods of microscopy (eg for biological specimens — the fibrous elements, flagellae, cilia, etc). Consequently, when separate phases are detected their Raman spectra can be recorded to distinguish chemical differences in their composition.

Figure 6.17 shows DIC photomicrographs of a cetostearyl alcohol/cetrimide emulsion/gel showing the different phases present and their associated Raman spectra.

Useful texts on Microscopy are 'Microscope Technique'. By W. Burrells and 'Hartley's Microscopy' by W. G. Hartley.

6.9 Temperature and Pressure Techniques with the Raman Microprobe

6.9.1 Temperature Cells

Controlled temperature (hot and cold) and controlled atmosphere sample stages are commercially available for optical microscopes which allow the study of solid phase transformations under controlled conditions (ie under air or inert atmosphere) on a small amount of sample. A suitable cell is shown in Fig. 6.18. This particular cell has a temperature range of $-196\,°C$ to $+600\,°C$, a similar cell has a temperature range of ambient to $1500\,°C$.

The sample to be studied is placed on the sample platform within the cell and an ultra long working distance objective used to view and record Raman spectra at various temperatures to follow the chemical transformations. A similar cell has been reported [14] and used to study phase transformation and hydrodesulfurization catalysts.

6.9.2 Pressure Measurements

Raman data obtained on samples at elevated pressures using a diamond anvil cell on a Raman microprobe were first reported by Gardiner et al. [15]. They reported results on the effect of pressure on the C—H stretching vibrations of decane and a commercial traction fluid. The sample was placed in the DAC along with a small ($\sim 50\,\mu m$) chip of ruby. The ruby is used as a measure of pressure within the cell. The shift in the ruby Raman spectrum relative to pressure is well documented.

The ruby chip is located within the DAC and its spectrum recorded, the position of the peaks giving a measurement of pressure within the DAC. The sample spectrum is then recorded at the same pressure. The pressure is then increased and the spectra of the ruby chip and samples are recorded at each pressure change.

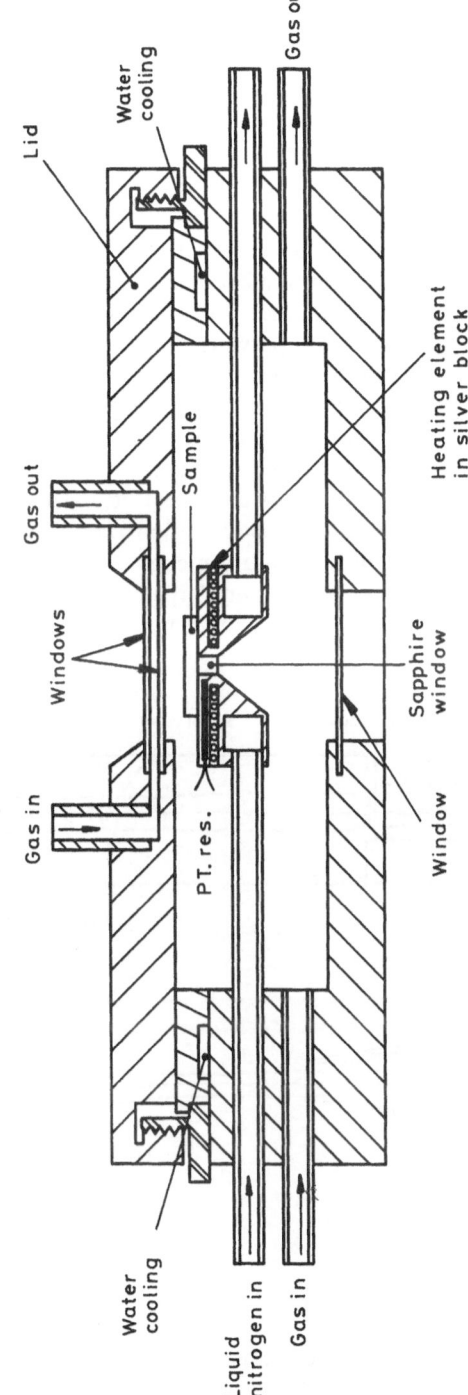

Fig. 6.18. Controlled temperature and atmosphere cell for use with the Raman microprobe. Reproduced by permission of Linkam Scientific Instruments

The Raman microprobe has also been used to measure the stress in the diamonds in the DAC [16]. Spectra of the diamonds under pressure were measured from selected volumes in both 180° and 135° scattering geometries. The diamond peak was found to shift linearly with pressure.

6.10 Polarisation Ratios

Information about the symmetry of the molecular vibrational mode giving rise to a Raman band can be obtained by the measurement of its polarisation properties [17]. The Raman microprobe would seem to be the ideal tool to obtain polarisation measurements from small single crystals, individual polymer filaments or heterogeneities in polymer films. However, due to the optical design of microspectrometers incorporating a dielectric beamsplitter the measurements are not as straightforward as would at first appear. Corrections must be made for any beamsplitter effects. Depolarisation due to the large collection angles of high magnification, high numerical aperture objectives is usually assumed to be small and is often neglected. Theoretical analysis [18] and practical measurement [19, 20] of depolarisation ratios has recently been reported. The experimental arrangement to measure depolarisation ratios is shown in Fig. 6.19, and Fig. 6.20 shows the parallel and perpendicular spectra of carbon tetrachloride reference liquid. Polarisation measurements are discussed in detail in Chapter 3.

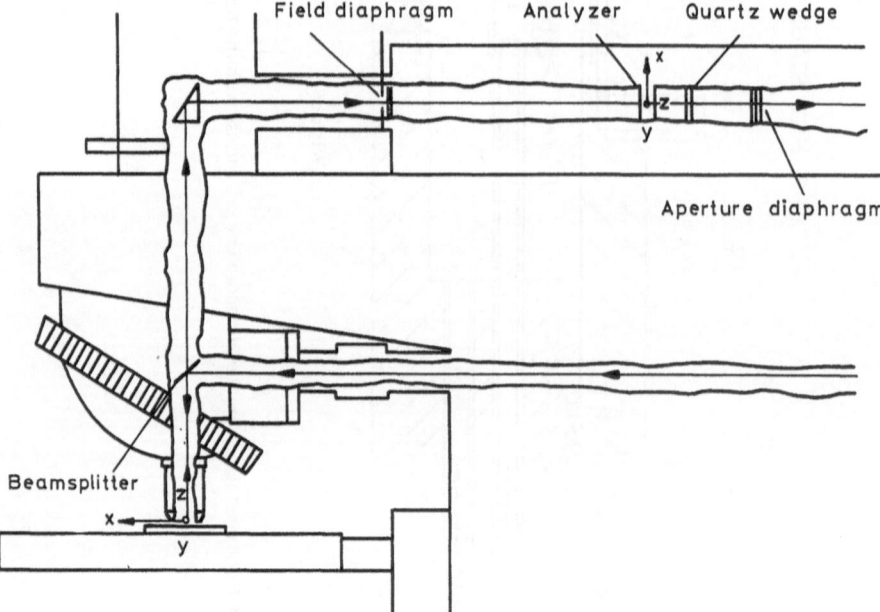

Fig. 6.19. Schematic of illumination and collection optical paths for polarisation studies (from Ref. 8). (Reprinted with permission from Analytical Chemistry (1981) 53: 1772. Copyright 1981 American Chemical Society)

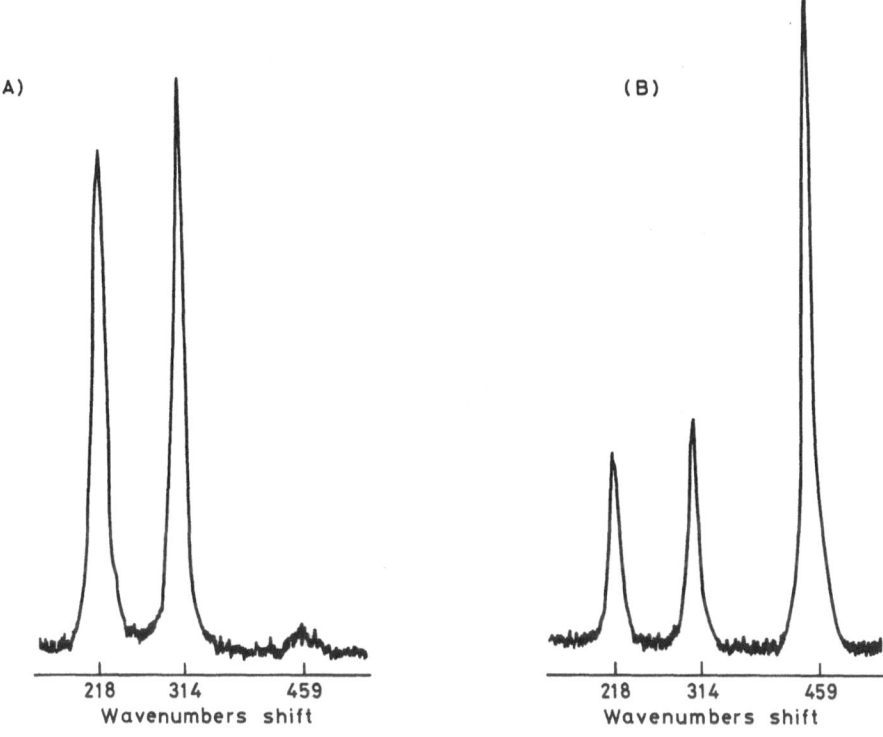

(A)

(B)

218   314   459
Wavenumbers shift

218   314   459
Wavenumbers shift

Fig. 6.20. Polarised Raman spectra of carbon tetrachloride. A) Z(XY)Z; B) Z(YY)Z

6.11 Raman Imaging and Mapping

A unique ability of the original Molecular Optical Laser Examiner, the MOLE, was the ability to obtain a microscopic image of a sample through the use of discrete wavelength segments of the spectrum of Raman scattered radiation originating from a laser-irradiated sample.

The system on the MOLE utilised the multiplex advantage of a two dimensional low light level television-type electronic camera detector.

The principle involved was that the sample was placed on the microscope slide and the laser beam illuminated the whole field of view of the microscope. This was achieved by rotating the laser beam to form an annular beam of light which was directed to the condenser of a dark-field microscope objective (Fig. 6.14). Thus the sample was uniformly irradiated with monochromatic radiation. The microscope then transferred the image of the sample to a monochromator which selected the desired wavelength to be imaged onto a television camera placed at the exit focal plane. If the wavelength selected corresponds to a Raman band frequency of a compound in the sample then the micrographic image indicated the distribution of this component throughout the illuminated area. The spatial resolution of the system was about 1 micron. This system worked well for samples which were good Raman scatterers but tended not to work with poor Raman scatterers and could not easily discriminate Raman features from fluorescent backgrounds. Figure 6.21 shows the principles of Raman imaging. The big advantage of this method of Raman imaging/ mapping was that the detector observed the entire field simultaneously.

Another approach to Raman mapping is to use a computer-controlled stepping motor driven sample stage on the microscope. In this procedure Raman spectra are detected at several hundred positions on a sample surface. A computer is programmed to record a spectrum on the first location on the sample and the spectrum stored on disk. The computer then transmits an instruction to the sample stage microcomputer to move to the next sampling point and the Raman spectrum recorded and stored. This procedure is repeated until all of the necessary data is recorded and stored.

In this way either one dimensional mapping line profiles or 2 dimensional Raman maps of the species present on the surface are built up. In this procedure it is the sample which is moved relative to the stationary focused laser beam [21].

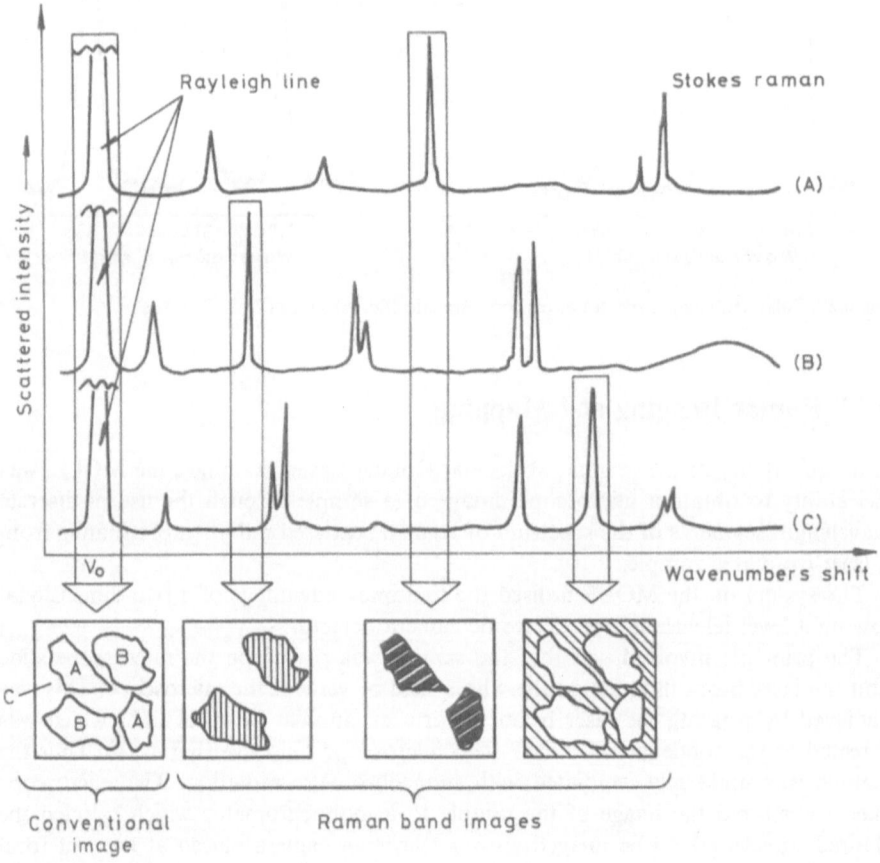

Fig. 6.21. Illustration of the principle of selective imaging by use of specific spectral bands of the Raman scattered radiation. (Reprinted with permission from John Wiley & Sons Ltd. GJ Rosasco, Raman microscope spectroscopy, in: Advances in Infra-Red and Raman Spectroscopy Vol. 7. Copyright Heyden & Son Ltd 1980)

A different approach is to scan the focused laser beam over the stationary sample and record the spectrum at different locations on the sample. This method [22] is achieved by a new kind of transfer optics placed between the microscope and the spectrometer, which enables an optimised coupling. The coupling optics consist of a pair of lenses. One lens focuses the laser beam into the back focal plane of the microscope objective lens. This lens, optically coupled to the back aperture of the objective can be moved in two orthogonal directions perpendicular to the laser beam. Thus this lens can focus the laser beam on any point in the microscope field and therefore on any point of the sample. The second lens is mechanically coupled to the first and placed in the scattered beam to balance any shift in the image probed area and focus the scattered beam on the entrance slit of the spectrometer. The optical arrangement has proved to be convenient for a variety of cumbersome samples such as cryostats, variable temperature or pressure cells.

6.12 Application Areas

6.12.1 Industrial Applications

In polymer film and fibre production any contamination within the films or fibres cause the appearance and properties to be impaired.

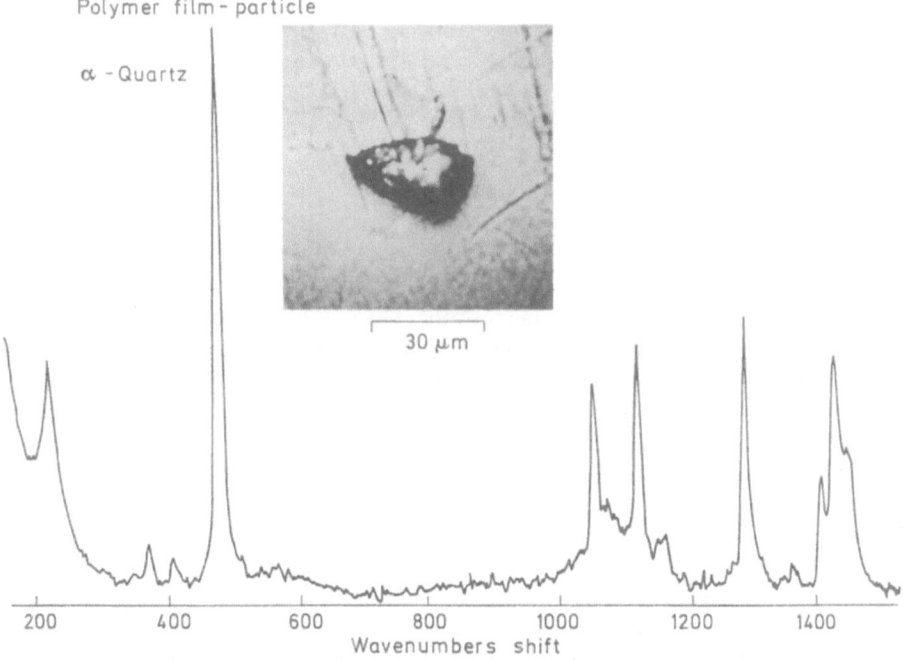

Fig. 6.22. Particulate contamination in polymer film. Peak at 465 cm^{-1} indicates α-SiO$_2$. Peaks at $\sim$1060, 1150, 1295 and 1450 cm^{-1} — polythene

Solid deposit in glass – Sodium sulphate

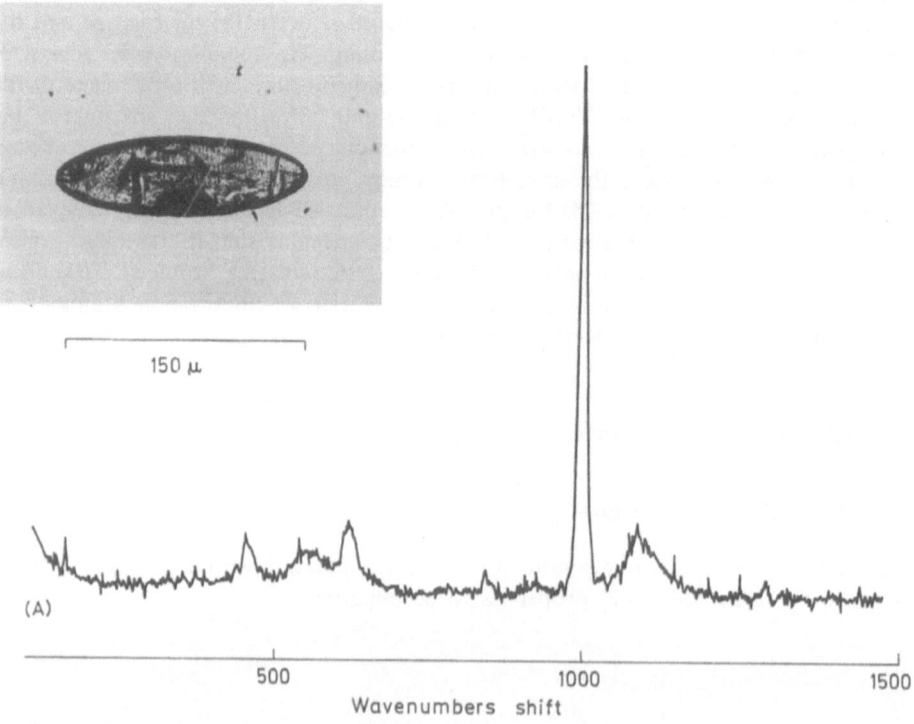

Fig. 6.23. Solid deposits in glass using ultra long working distance objectives. A Sodium sulphate

The identification of these impurities often enables the causes of the problem to be found and corrected. Raman microprobes have been successfully used to identify contaminants (Fig. 6.22), local differences in crystallinity and local concentrations of copolymers in polymer processes [23–25].

The manufacture of microcircuits involves a large number of complicated chemical operations, during which accidental contaminations may impair the functioning of the circuit. Impurities in semi-conductor crystals and integrated circuits have been characterised [26–28] and enable the causes of contamination to be pinpointed and eliminated.

Analysis of the inclusions and deposits in bubbles present in glass can yield valuable information about the refining mechanism or causes of unwanted bubble formation (Fig. 6.23). Other industrial applications include studies of crystallinity in graphite and carbon fibres [29, 30] and industrial catalysts to detect the degree of 'activation' achived in preparation of the catalyst [11]. Dopant concentration levels in optical fibres have been measured [31, 12].

Solid deposit in glass - Sulphur

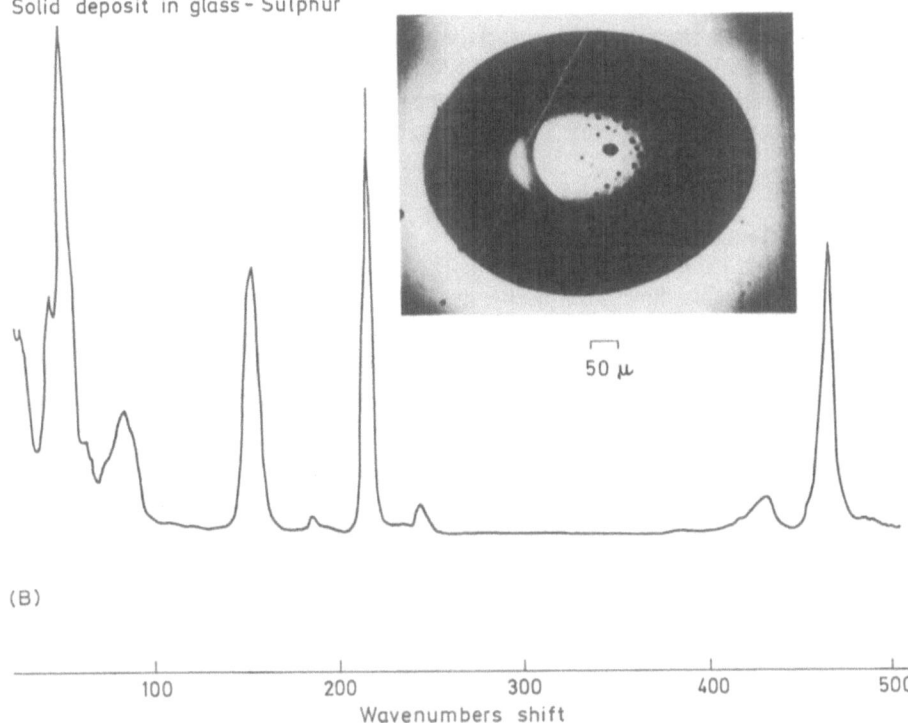

(B)

| 100 | 200 | 300 | 400 | 500 |

Wavenumbers shift

Fig. 6.23. B Sulphur

6.12.2 Biology and Pathology

In situ identification and localisation of concretions of uric acid, guanine, sodium and potassium urates and xanthene in tissues of fish, spiders, molluscs and insects [32, 33], the accumulation and composition of hydrocarbon in *Botryococcus braunii* [34], and accumulation of copper sulphide needles inside the Cytosomes of *Littorina littorea* [35], have all been reported. A successful area of application has been in the study of biological mineralisation in tooth and bone [36] and atherosclerotic and bioprosthetic calcification in man and animals [37], and characterisation of Bilirubin-Type gallstone [38].

6.12.3 Mineralogy and Geology

Many investigations of in situ analysis of fluid inclusions in minerals have been reported [39–46]. Fluid inclusions are multiphase systems — coexisting liquid, gaseous and solid phases — trapped within a mineral host. Analysis of the contents of the inclusion provides important data related to many mineralogical, geochemical geochemical and geological processes.

Inclusions in precious stone [47] have been identified and in a micropaleontological study the classification of Foraminifera has been solved by revealing the calcitic nature of one species attributed to an aragonitic family [48].

In the study of some sheet (talc) and ribbon (amphiboles) silicates it was found that bulk powder samples of high surface area varieties (platy and fibrous talc and fibrous amphiboles) did not yield useful Raman spectra whereas microprobe analysis of individual microscopic crystals yielded good quality spectra for all varieties studied [49]. Characterisation of inclusions in eclogites [50] and feldspars [51] have been reported.

6.12.4 Environmental Studies

Raman microprobes have been used extensively in the analysis of individual microscopic particles in the context of environmental pollution studies. Characterisation of airborne urban dusts [52, 53], power plant emissions [54, 55], and stratospheric aerosols collected at the South Pole [56] have been reported. A wide variety of inorganic

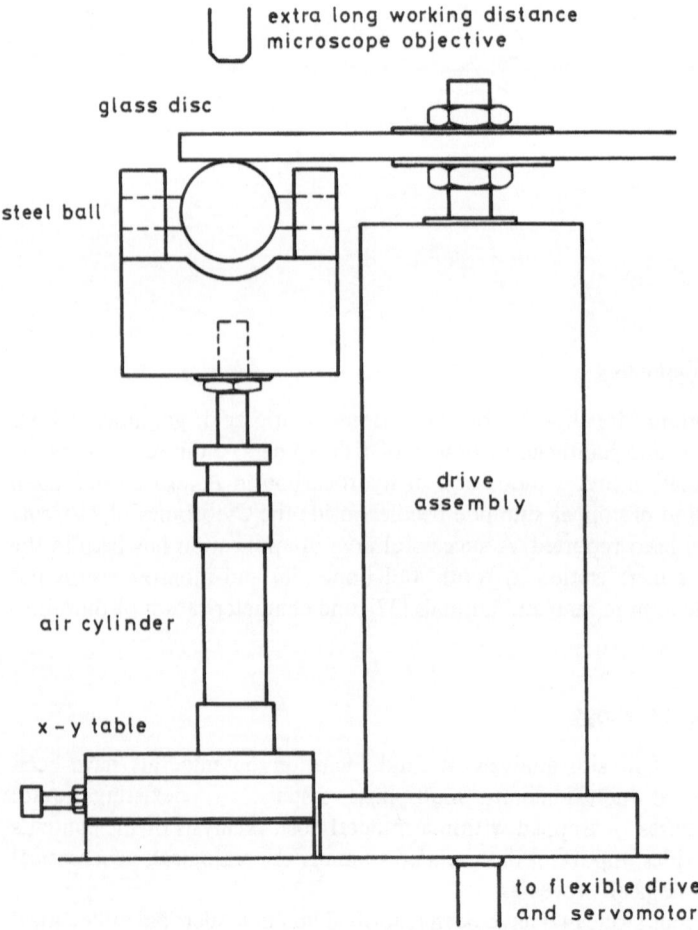

Fig. 6.24. Dynamic ball and plate Raman lubrication rig (from Ref. 31)

materials have been identified including quartz, calcite, dolomite, anhydrite, sodium nitrate, vanadium pentoxide, ammonium sulfates.

6.12.5 Corrosion Studies

Corrosion of metal surfaces involves complex reactions between the surface material and the surrounding environment. These reactions ultimately lead to metallurgical failure. Critical information about the corrosion process can be inferred from characterisation of the reaction products on the corroded surface and hence an understanding of the interactions between a material and its environment can be obtained. The Raman microprobe is proving itself to be a valuable tool in identifying corrosion compounds on metal surfaces.

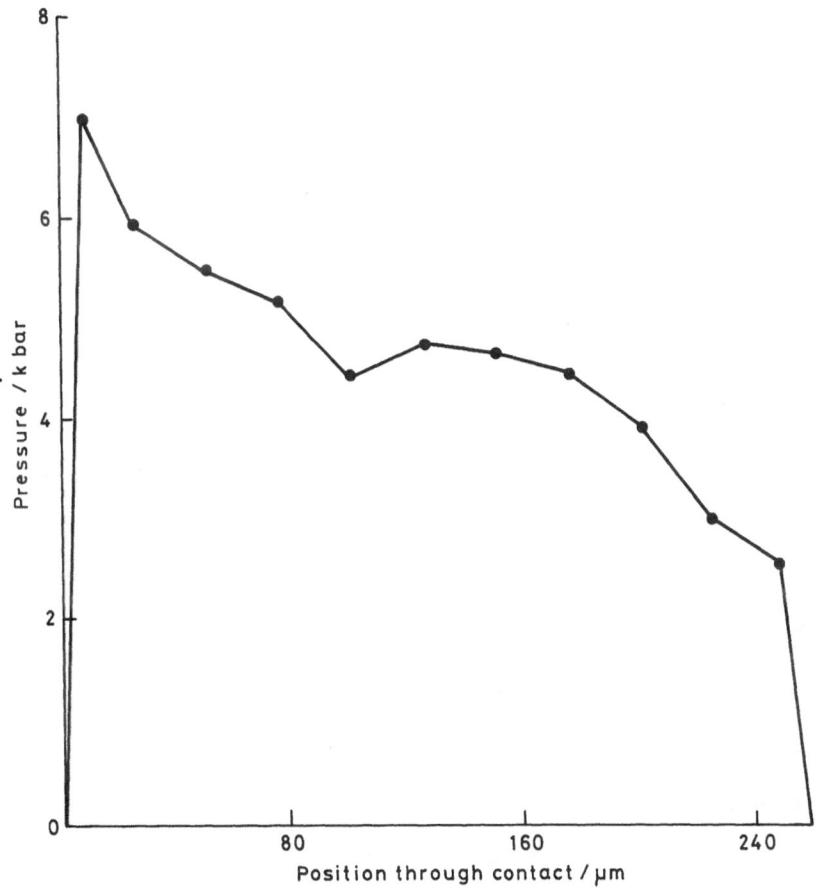

Fig. 6.25. Pressure profile through a rolling EHD contact, determined from Raman shift measurements. (260 µm represents entrance, 0 µm represents exit). (1 bar = 10^5 Pa) (from Ref. 31). (Reprinted with permission from Phil. Trans. R. Soc. Lond. (1986) A 320)

The corrosion products formed on electrodes subjected to corona discharges [57], the products formed on weathering steels [58], and products formed on molybdenum, hafnium, titanium, tantalum and zirconium in high sulphur low oxygen environment [12] have been studied.

Corrosion or corrosion-inhibition processes have been studied in situ on the Raman microprobe using a long working distance objective. The corrosion inhibition films formed on copper in acidic aqueous environments by alkyl substituted imidazoline-2-thione have been characterised by comparison of the Raman spectra from the copper surface to those obtained from copper complexes of the inhibitors prepared from simple copper salts [59].

6.12.6 Rolling Elastohydrodynamic Contacts

A novel application of Raman microscopy reported by Gardiner et al. [12] is the study of the pressure experienced by a lubricant in the contact between a steel ball and glass disc. In high-load bearing contacts, the bearing surfaces elastically deform around a thin film of the lubricant. These elastohydrodynamic contacts can be modelled by using a rolling steel ball on a glass plate which allows optical microscopic analysis of the contact region [60]. Using the Raman microprobe the Raman spectrum of the lubricant in the contact area could be obtained and monitored as a function of load, speed and position within the contact. The apparatus used and pressure profile obtained are shown in Fig. 6.24 and 6.25.

6.13 References

1. Delhaye M, Migeon M (1966) Compt. Rend. 262: 702
2. Hirschfeld T (1973) J. Opt. Soc. Am. 63: 476
3. Rosasco GJ, Etz ES, Cassatt WA (1974) IVth Internat. Conf. Raman Spectrosc., Brunswick, ME, USA
4. Delhaye M, Dhamelincourt P (1974) IVth Internat. Conf. Raman Spectrosc., Brunswick, ME, USA
5. Rosasco GJ, Etz ES, Cassatt WA (1975) Applied Spectroscopy 29: 396
6. Cook BW, Louden JD (1979) J. Raman Spectrosc. 8: 249
7. Dhamelincourt P (1982) Microbeam Analysis, p 261
8. Rosasco GJ, Etz ES (1977) Res. Devel. 28: 20
9. Andersen ME, Muggli RZ (1981) Anal. Chem. 53: 1772
10. Dhamelincourt P (1979) PhD thesis, University of Lille
11. Adar F, Clarke DR (1982) Microbeam Analysis, p 307
12. Gardiner DJ, Bowden M, Graves PR (1986) Phil Trans R Soc Lond A 320
13. Melveger AJ (1972) J. Polymer Sci., part A-2, vol 10, p 317
14. Payen E, Dhamelincourt MC, Dhamelincourt P, Grimblot J, Bonnelle JP (1982) App. Spectros. 36: 1
15. Gardiner DJ, Bowden M, Daymond J, Gorvin AC, Dare-Edwards MP (1984) Appl. Spectros. 38: 2
16. Sharma SK, Mao HK, Bell PM, Xu JA (1985) J. Raman Spectrosc. 16: 5
17. Damen TC, Porto SPS, Tell B (1966) Phys. Rev. 142: 570
18. Turrell G (1984) J Raman Spectrosc. 15: 2
19. Bremard C, Dhamelincourt P, Laureyns J, Turrell G (1985) Appl. Spectrosc. 39: 6

20. Andersen ME, Delly JG (1985) Microbeam Analysis, p 33
21. Gardiner DJ, Littleton CJ, Bowden M (1987) Microbeam Analysis, p 131
22. Barbillat J, Delhaye M (1983) Microbeam Analysis, p 280
23. Ogilvie GD, Addyman L: L'actualité Chimique Avril 1980: 51
24. Dhamelincourt P et al. (1979) Anal Chem 51: 414A
25. Delhaye M, Dhamelincourt P, Wallart F (1979) Toxic and Envir Chem Reviews 3: 73
26. Needham C: L'actualité Chimique Avril 1980 43
27. Ramsey JN (1984) J De Physique C2 2 45
28. Lang P, Katon JE (1986) Microbeam Analysis, p 47
29. Adar F, SEM/1979/I SEM Inc, O'Hare AMF IL 60666 1979.
30. Johnson WL, III (1986) Microbeam Analysis, p 26
31. Carvalho W, Dumas P (1984) J De Physique C2, 2 45, p 765
32. Truchet M et al.: L'actualité Chimique Avril 1980: 15
33. Dhamelincourt P (1979) Microbeam Analysis, p 155
34. Largeau C et al. (1980) Phytochem. 19: 1043
35. Martoja M, Tue VT, Elkaim B (1980) J. Exp. Mar. Biol. Ecol. 43: 251
36. Casciani FS, Etz ES (1979) Microbeam Analysis, p 169
37. Etz ES, Tomazic BB, Brown WE (1986) Microbeam Analysis, p 39
38. Zheng S, Tu AT (1986) Appl. Spectrosc. 40: 8
39. Rosasco GJ, Roedder E (1976) Internat. Geol. Congr. 25th Abstr. 3: 812
40. Dhamelincourt P, Schubnel HJ (1977) Rev Gemnol 52: 11
41. Debussy J, Dhamelincourt P, Poty B (1978) Colloque Mineraux et Minerais, Abstr, Nancy, France
42. Guilhaumou N, Dhamelincourt P, Touray JC, Barbillat J (1978) C R Acad Sci Ser D 287(15): 1317
43. Rosasco GJ, Roeder E (1979) Geochim Cosmochim Acta 43: 1907
44. Debussy J, Beny C, Guilhaumou N, Dhamelincourt P, Poty B (1984) J De Physique. C2 2: 45, p 811
45. Pasteris JD, Seitz JC, Wopenka B (1985) Microbeam Analysis, p 25
46. Higgins KL, Stein CL (1986) Microbeam Analysis, p 31
47. Dele-Dubois ML, Dhamelincourt P, Schubnel HJ: L'actualité Chimique Avril 1980, p 39
48. Venec-Peyre MT, Jaeschhe-Boyer H: L'actualité Chimique Avril 1980, p 32
49. Blaha JJ, Rosasco GJ (1978) Anal. Chem. 50: 892
50. Boyer H, Smith DC (1984) Microbeam Analysis, p 107
51. Purcell FJ, White WB (1983) Microbeam Analysis, p 289
52. Dhamelincourt P et al. (1979) Anal Chem 51: 414A
53. Etz ES, Rosasco GJ, Cunningham WC (1977) Environ analy, Academic Press, New York
54. Etz ES, Rosasco GJ (1977) Dimensions/NBS 61: 22
55. Doyle TE, Alvarez JL (1983) Microbeam Analysis, p 277
56. Cunningham WC, Etz ES, Zoller WH (1979) Microbeam Analysis, p 148
57. Le Ny R, Fiaud C, Nguyen AT (1984) J de Physique C2 2 45, p 661
58. Heidersbach R, Purcell F (1984) Microbeam Analysis, p 61
59. Gardiner DJ, Gorvin AC, Gutteridge C, Raper ES (1985) Corrosion Sci 25: 1019
60. Foord CA, Wedeven LD, Westlake FJ, Cameron A (1970) Proc. Instn. Mech. Engrs. 184: 487

CHAPTER 7

Further Reading

In this final section of the book, it is intended to refer the interested reader to review articles and texts which give a fuller treatment of areas of either experimental practice or theory. Some of these areas will have been introduced in the previous chapters, others represent aspects of Raman spectroscopy not previously discussed. Non-linear, multi-photon techniques, for example, are being developed by many research groups but have not been included in the main body of the text. A better grasp of fundamental theory is an important aim of any experimentalist, therefore references to molecular dynamics and scattering theory are also included in this section. Neither SERS nor resonance-Raman experiments were mentioned, reviews of these subjects have been published by Chang and Furtak and by Yamada, respectively. The use of circularly-polarized light was briefly treated in Chap. 2, Sect. 12, although the Raman spectroscopy of chiral molecules was not considered. The appropriate theory has been reviewed by Barron, while experimental aspects are discussed in the article by Kiefer.

Time-resolved Raman spectroscopy, a subject which involves very specialised and difficult practical techniques, has been treated by Bridoux and Delhaye and more recently by Atkinson. An excellent general review of Raman sampling devices, including double-beam techniques is that by Kiefer. Finally, band-shape analysis and its application in molecular dynamics was mentioned only in passing in previous chapters. The interested reader is refered to the review by Clark and the recent book by Rothschild for a complete development of this interesting area of Raman spectroscopy.

Band shape analysis

Clark RJH (1978) Band shapes and molecular dynamics in liquids. In: Clark RJH, Hester RE (eds) Advances in infrared and Raman spectroscopy, vol 4. Heyden, London

Group theory and aspects of Raman scattering in solids

Turrell G (1972) Infrared and Raman spectra of crystals. Academic Press, London
Sherwood PMA (1972) Vibrational spectroscopy of solids. In: Linnett JW, Purnell JH (eds) Cambridge monographs in physical chemistry 1, Cambridge University Press, Cambridge

Molecular dynamics

Rothschild WG (1984) Dynamics of molecular liquids. Wiley Interscience, New York
Wilson EB Jr, Decius JC, Cross PC (1955) Molecular vibrations. McGraw-Hill, New York
Gordon RG (1965) J. Chem. Phys. 43: 1307

Non-linear Raman techniques

Eesley GL (1981) Coherent Raman spectroscopy. Pergamon Press, Oxford

Raman circular dichroism

Kiefer W (1977) Recent techniques in Raman spectroscopy. In: Clark RJH, Hester RE (eds) Advances in infrared and Raman spectroscopy, vol 3. Heyden, London
Barron LD (1978) Raman optical activity. In: Clark RJH, Hester RE (eds) Advances in infrared and Raman spectroscopy, vol 4. Heyden, London

Resonance Raman scattering

Behringer J (1974) Theories of resonance Raman scattering. In: Molecular Spectroscopy, vol 2
Johnson BB, Peticolas WL (1976) Ann. Rev. Phys. Chem. 27: 465

Statistical analysis of data

Chatfield C (1983) Statistics for technology (3rd edition). Chapman and Hall, London and New York
Chatfield C (1980) The analysis of time series (2nd edition). Chapman and Hall, London and New York

Surface enhanced Raman scattering

Chang RK, Furtak R (eds) (1982) Surface enhanced Raman scattering. Plenum Press, New York
Yamada H (1981) Appl. Spectros. Rev. 17: 227

Time-resolved Raman spectroscopy

Bridoux M, Delhaye M (1976) Time-resolved and space-resolved Raman spectroscopy. In: Clark RJH, Hester RE (eds) Advances in infrared and Raman spectroscopy, vol 2. Heyden, London
Atkinson GH (1982) Time-resolved spectroscopy. In: Clark RJH, Hester RE (eds) Advances in infrared and Raman spectroscopy, vol 9. Heyden, London

Subject Index

D. L. Andrews

Lasers in Chemistry

1986. 115 figures. XII, 176 pages.
ISBN 3-540-16161-9

Contents: Principles of Laser Operation. – Laser
Sources. – Laser Instrumentation in Chemistry. –
Chemical Spectroscopy with Lasers. – Laser-
Induced Chemistry. – Appendix 1: Listing of
Output Wavelengths from Commercial Lasers. –
Appendix 2: Directory of Acronyms and Abbrevia-
tions. – Appendix 3: Selected Bibliography. –
Subject Index.

This book gives a concise overview of the subject
accessible to a more general readership, and whilst
the emphasis is placed on chemical topics, applications in other fields are also represented. The
level of background knowledge assumed here
corresponds roughly to the material covered in a
first-year undergraduate course in chemistry. There
has been a deliberate effort to avoid heavy mathe-
matics.
The link between lasers and chemistry is essentially a three-fold one. Firstly, several important
chemical principles are involved in the operation of
most lasers. Secondly, a large number of techniques based on laser instrumentation are used to
probe systems of chemical interest; and the third
link between lasers and chemistry concerns the
inducing of chemical change in a system through
its irradiation with laser light. This book provides
examples illustrating the diversity of laser applications in chemistry across the breadth of the scientific spectrum from fundamental research to
routine chemical analysis. Nonetheless the emphasis is mostly placed on applications which have
relevance to chemical industry.

Springer-Verlag Berlin
Heidelberg New York London
Paris Tokyo Hong Kong